Engineering Mechanics:
Dynamics

Engineering Mechanics: Dynamics

Contributors
_____

Gabriel Barceló et al.

AURIS
Reference

www.aurisreference.com

Engineering Mechanics: Dynamics

Contributors: Gabriel Barceló et al.

Published by Auris Reference Limited

www.aurisreference.com

United Kingdom

Engineering Mechanics: Dynamics

ISBN: 978-1-78154-816-5

British Library Cataloguing in Publication Data
A CIP record for this book is available from the British Library

Printed in the United Kingdom

Exclusively distributed by CBS Publishers & Distributors Pvt. Ltd.

Sales & Distribution Rights only for India, Pakistan, Bangladesh, Sri Lanka, Nepal and Bhutan.This book is not to be sold outside these territories.

Contents

List of Abbreviations

ATMDs	Active tuned mass dampers
BEM	Boundary element method
CISS	Critical interfacial shear stress
CNT	Carbon nanotube
DOFs	Degrees of freedom
DR	Dead reckoning
FEM	Finite element modeling
GPS	Global Positioning System
GTR	General theory of relativity
HFBB	High-frequency base balance
IMU	Inertial Measurement Unit
ISS	Interfacial shear stress
MAV	Micro air vehicle
MCS	Monte carlo simulation
MD	Molecular dynamics
OMA	Optical-mechanical analogy
PDF	Probability density function
PIV	Particle image velocimetry
PS	Polystyrene
RBFs	Radial basis functions
SDOF	Single degree of freedom
SFE	Stochastic finite element
SIFT	Scale-Invariant Feature Transform
SISS	Steady sliding interfacial shear stress
SLS	Standard linear solid
SNR	Signal-to-noise ratio
SSFEM	Spectral stochastic finite element method
SWCNT	Single-walled CNT
TMDS	Tuned mass dampers
UIF	Unscented Information Filter
WVU	West Virginia University

List of Contributors

Gabriel Barceló
Advanced Dynamics S.A., Madrid, Spain

Yoshihiro Kubota
Faculty of Science and Engineering, Toyo University, Saitama, Japan

Osamu Mochizuki
Faculty of Science and Engineering, Toyo University, Saitama, Japan

Chen Liu
Department of Transportation and Environmental Engineering, Graduate School of Engineering Science, Hiroshima University, Higashi-Hiroshima, Japan

Yoshikazu Tanaka
Department of Transportation and Environmental Engineering, Graduate School of Engineering Science, Hiroshima University, Higashi-Hiroshima, Japan

Yukio Fujimoto
Department of Transportation and Environmental Engineering, Graduate School of Engineering Science, Hiroshima University, Higashi-Hiroshima, Japan

Javed I. Siddique
Department of Mathematics, Pennsylvania State University, York Campus, York, USA

Forrest A. Landis
Department of Chemistry, Pennsylvania State University, York Campus, York, USA

Muhammad R. Mohyuddin
Department of Mathematics, FAST University, Islamabad, Pakistan
NCBA & E, Gujrat, Pakistan

Alfredo Bacchieri
University of Bologna, Bologna, Italy

Diego A. Carranza
Instituto de Astronoma, Universidad Nacional Autónoma de México, México D.F., México

Sergio Mendoza
Instituto de Astronoma, Universidad Nacional Autónoma de México, México D.F., México

Arbab Ibrahim Arbab
Department of Physics, Faculty of Science, University of Khartoum, Khartoum, Sudan

E. Goulart
Instituto de Cosmologia Relatividade Astrofisica (ICRA-Brasil/CBPF) Rua Dr. Xavier Sigaud, 150, 22290-180 Rio de Janeiro, RJ, Brazil

Michele Betti
Department of Civil and Environmental Engineering (DICeA), University of Florence, Street Santa Marta 3, 50139 Florence, Italy

Paolo Biagini
Department of Civil and Environmental Engineering (DICeA), University of Florence, Street Santa Marta 3, 50139 Florence, Italy

Luca Facchini
Department of Civil and Environmental Engineering (DICeA), University of Florence, Street Santa Marta 3, 50139 Florence, Italy

T. G. Ritto
PUC-Rio Department of Mechanical Engineering 22453-900 Rio de Janeiro, RJ, Brazil. Université Paris-

Est Lab. de Modélisation et Simulation Multi-Echelle 77454 Marne-la-Vallée, France

C. Soize
Université Paris-Est Lab. de Modélisation et Simulation Multi-Echelle 77454 Marne-la-Vallée, France

Rubens Sampaio
PUC-Rio Department of Mechanical Engineering 22453-900 Rio de Janeiro, RJ, Brazil

Aly Mousaad Aly
Department of Mechanical Engineering, Faculty of Engineering, Alexandria University, Alexandria 21544, Egypt

Alberto Zasso
Department of Mechanical Engineering, Politecnico di Milano, Via G. La Masa 1, 20156 Milano, Italy

Ferruccio Resta
Department of Mechanical Engineering, Politecnico di Milano, Via G. La Masa 1, 20156 Milano, Italy

Z. Q. Zhang
Center of Micro/Nano Science and Technology, Jiangsu University, Zhenjiang 212013, China
Department of Engineering Mechanics and State Key Laboratory of Structural Analysis for Industrial Equipment, Dalian University of Technology, Liaoning, Dalian 116024, China

D. K. Ward
Radiation and Nucleation Detection Materials and Analysis Department, Sandia National Laboratories, Livermore, CA 94550, USA

Y. Xue
Department of Mechanical and Aerospace Engineering, Utah State University, Logan, UT 84322, USA

H. W. Zhang
Department of Engineering Mechanics and State Key Laboratory of Structural Analysis for Industrial Equipment, Dalian University of Technology, Liaoning, Dalian 116024, China

M. F. Horstemeyer
Center for Advanced Vehicular Systems, Mississippi State University, Mississippi State, MS 39762, USA

Matthew B. Rhudy
Division of Engineering, Pennsylvania State University, Reading, PA 19610, USA

Yu Gu
Department of Mechanical and Aerospace Engineering and Lane Department

of Computer Science and Electrical Engineering at WVU, Morgantown, WV
26506, USA

Haiyang Chao
Aerospace Engineering Department, University of Kansas, Lawrence, KS
66045, USA

Jason N. Gross
Department of Mechanical and Aerospace Engineering at West Virginia University, Morgantown, WV 26506, USA

Preface

Engineering mechanics is the application of mechanics to solve problems involving common engineering elements. The goal of engineering mechanics course is to expose problems in mechanics as applied to plausibly real-world scenarios. The text Engineering Mechanics: Dynamics provides a solid foundation of mechanics principles. First chapter focuses on analysis of dynamics fields in noninertial systems. The purpose of second chapter is to understand the appropriate shape of plate, and the role of vortex to fly or glide the longer distance with a rotating thin plate. In third chapter, we present viscosity transient phenomenon during drop impact testing and its simple dynamics model. In fourth chapter, we explore the one-dimensional drainage of a power-law fluid into a deformable porous material. The light as composed of longitudinal-extended elastic particles obeying to the laws of Newtonian mechanics has been discussed in fifth chapter. Modified Newtonian dynamics as an entropic force has been presented in sixth chapter. The generalized Newton's law of gravitation versus the general theory of relativity has been outlined in seventh chapter. The aim of eighth chapter is to show that Newtonian mechanics of point particles in static potentials admits an alternative description in terms of effective Riemannian spacetimes. Ninth chapter presents a hybrid Galerkin/perturbation approach based on radial basis functions for the dynamic analysis of mechanical systems affected by randomness both in their parameters and loads. Stochastic dynamics of a drill-string with uncertain weight-on-hook has been proposed in tenth chapter. Eleventh chapter presents a procedure for the response prediction and reduction in high-rise buildings under multidirectional wind loads. Twelfth chapter examines the CNT/polymer interfacial interactions to determine the dominant mechanisms responsible for load transfer using molecular dynamics (MD) simulations. Last chapter offers a set of novel navigation techniques that rely on the use of inertial sensors and wide-field optical flow information.

Chapter 1

ANALYSIS OF DYNAMICS FIELDS IN NONINERTIAL SYSTEMS

Gabriel Barceló
Advanced Dynamics S.A., Madrid, Spain

ABSTRACT

In this paper, I present evidence that there exists an unstructured area in the present general assumptions of classical mechanics, especially in case of rigid bodies exposed to simultaneous noncoaxial rotations. To address this, I propose dynamics hypotheses that lead to interesting results and numerous noteworthy scientific and technological applications. I constructed a new mathematical model in rotational field dynamics, and through this model, results based on a rational interpretation of the superposition of motions caused by torques were obtained. For this purpose, I analyze velocity and acceleration fields that are generated in an object with intrinsic angular momentum, and assessed new criteria for coupling velocities. In this context, I will discuss reactions and inertial fields that cannot be explained by classical mechanics. The experiments have been analyzed and explained in a video accompanying this text. I am not aware of any concurrent study on the subject and conclusions evidenced in this paper, preventing us from making additional theoretical comparisons or indicate to the reader other sources to compare criteria.

INTRODUCTION

To date, no mathematical correlation exists in the laws of mechanics that relates orbital and rotational motions, despite this effect being ubiquitous in nature. I, therefore, investigated whether a physical correlation exists between these motions, and if it does, then the derivation of its mathematical expression.

In this study, I investigated noninertial systems to better understand the response of rigid bodies subjected to simultaneous noncoaxial rotations. As a result, I proposed hypotheses that require extending my studies to field theory in order to explain the dynamics of such bodies.

The superposition of simultaneous motions was discussed by Galileo [1] to explain the path of a canyon ball. Understanding the superposition of fields caused by torques is important, and I shall refer to the superposition of the velocity fields generated in these cases as "coupling."

I analyzed the motion of objects subjected to multiple rotations, such as a spinning top, a boomerang, a hoop, and several other objects whose peculiar behavior is intriguing. I believe that my conclusions modify the foundations of rational dynamics and incorporate new criteria of great impact and significance.

EXPERIMENTS AND MATHEMATICAL SIMULATION

I investigated how objects respond when subjected to different simultaneous accelerations by applying noncoaxial rotations. In other words, I investigated the motion of a gyroscope, spinning top, or boomerang to determine whether it can be described by the classical mechanics or it corresponds to laws of nature not yet formulated.

To clarify these questions, I performed several experiments that eventually indicate that mathematical formulation should use field theory and consider the concept of dynamics interactions. This implies that the generated dynamics fields interact to cause paths that correspond to some laws of physics that are not yet formulated.

My initial experiments motivated me to establish a dynamics hypothesis stating that under conditions of simultaneous noncoaxial rotations, an object follows a closed path (i.e., orbital motion) without requiring a centripetal force [2,3] and maintains its rotation about its initial axis. By numerical simulation, I confirmed that an object traces a closed path similar to the orbit of a body in space subjected to a centripetal force. However, no such centripetal force existed in my simulation.

Therefore, I was confronted with the challenge of reproducing such a path experimentally.

Watch the video by clicking on the Figure 1 or in this direction: http://dl.dropbox.com/u/48524938/VTS_Ingles.mov

FINAL TESTS: PROTOTYPE I

Because performing tests in space is difficult and expensive, my tests were conducted in water. I designed a cylinder or "submarine" driven by a stern-mounted propeller that translated and simultaneously forced the prototype to rotate about its longitudinal axis. The prototype also experienced torque because of its weight and upward buoyancy, both of which acted perpendicular to its longitudinal axis (i.e., its axis of rotation). I experimented using two different prototypes: one with fixed torque (prototype I) and another with variable torque (prototype II; see next section).

Figure 1: Theory of dynamic interactions: video illustrating experimental tests.

For a flat disk rotating about its axis of symmetry (Figure 2), velocity fields due to the rotation of the disk can be identified. For each portion of the disk, the tangential velocity is given by the following equation:

$$\boldsymbol{v} = \boldsymbol{\delta} \times \boldsymbol{\omega} \tag{1}$$

with

δ: Vector with distance from the particle to the rotation axis.

ω: Rotation speed of the disc.

Note that d is also the circumference ratio (Figure 2) or the geometric position of the particles whose dynamic state is being analyzed and that are equidistant from the rotation axis.

Therefore, all particles situated at the given radial position have the same tangential velocity but move in different directions, as already demonstrated E. Jouffret [4- 6]. Thus, turning the disk about its axis of symmetry creates a homogeneous and balanced field of velocities (Figure 2).

Prototype I experienced constant-torsion torque because the forces of weight and buoyancy act on different points to generate an imbalance. This torque identified by M', which I call the interaction torque, forced the submerged cylinder to turn about a new axis, i.e., the torque axis. However, in line with my hypothesis, I calculated that the cylinder cannot undergo this new turn but can trace a closed circular path.

The velocity of prototype I was remote controlled, and it did not have a rudder or any other steering device. When prototype I was at rest, the torque produced by gravity made the stern sink because prototype I was not balanced.

It reached a vertical position with the bow slightly above the water surface. However, once the motor was started and prototype I began rotating on its longitudinal axis, it turned right and initiated a closed path on the water surface with a radius that varied with the translational velocity.

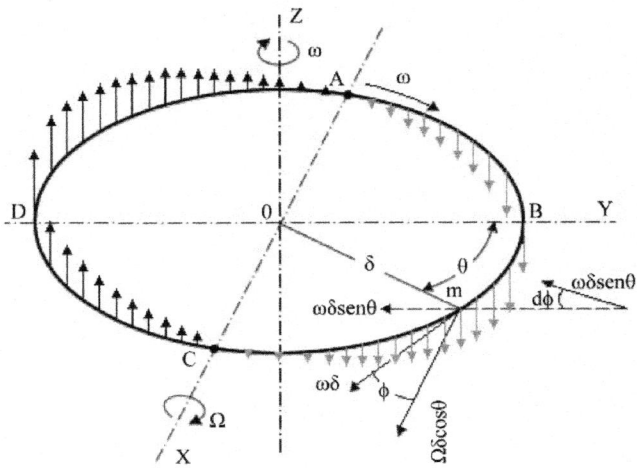

Figure 2: Body with angular velocity ω about its principal axis. When the body is subjected to a noncoaxial angular velocity Ω, a nonhomogeneous field of velocities is generated.

These tests clearly confirmed that prototype I behaved as predicted by the simulation. With no rudder but equipped with its own source of torque about its longitudinal axis, prototype I transformed this torque into an orbital motion, dynamically adding or coupling the linear translational velocity with the centripetal velocity produced by the interaction torque. The resulting path of prototype I was circular despite the absence of external central force that generates such a path. The circular orbit transformed into an elliptical orbit upon modifying the velocity of prototype I. As predicted by my theory, there was clear cause and effect.

The observed behavior can be understood with the help of the field theory mentioned above. Let us analyze the velocity fields that are generated in a section of the prototype. The torque generated by weight and buoyancy generates a second rotation noncoaxial with the existing one on the prototype's longitudinal axis. This second rotation determines anisotropic fields of velocity and acceleration (see Figure 1 video). However, something exceptional occurs: this new initial velocity distribution is modified when the prototype turns about its longitudinal axis.

When the prototype is turning about its longitudinal symmetry axis, a new velocity distribution is obtained, as seen in the Figure 1 video. The result is that, in line with the analysis of the velocity field, the torque produced by weight and buoyancy generates a new nonhomogeneous velocity field, which rotates the body about an axis different from that about which the external gravitational torque acts. The result is the observed circular path.

Thus, the rotation generated by the gravitational interaction torque does not correspond to the classical laws of mechanics or to vector algebra. In my opinion, after a half turn, the second rotation began about the axis perpendicular to the torque it generated, and not about the axis torque created by the weight and the buoyancy thrust. Once I determined this particular behavior of bodies subjected to simultaneous noncoaxial rotations, which it is defined in classical mechanics as the gyroscopic effect [7,8], I continued to analyze the real behavior of the proto-type.

The new velocity field created by the second rotation is not homogeneous, as is the case for the first rotation, but is an anisotropic velocity field. In line with my hypothesis, this new nonhomogeneous velocity field couples dynamically to the translational velocity field, which implies that both fields add algebraically, while the rotation about the longitudinal axis remains independent, as can be seen in the **Figure 1** video.

My main hypothesis is as follows: my experiments confirmed what I detected in the simulation, namely, the translational velocity field of the cylinder couples to the velocity field generated by the torque produced by weight and buoyancy. As a result, prototype I executed a circular path, as can also be observed in the flight of a boomerang or in the path of a spinning top.

PROTOTYPE II

After concluding the tests with prototype I, I modified it so that it could be steered to the port or starboard by exerting a variable external torque on it. This prototype, named prototype II, was equipped with a mechanism to exert a variable torque by using a pump and two water reservoirs (Figure 3). By transferring the water between the reservoirs, I modified the center of mass of prototype II, which enabled me to sink the bow or stern and thus to change the gravitational interaction torque.

The tests conducted with prototype II completely confirmed my theory. I steered the submarine to the port or starboard, in line with the variation of the gravitational torque. Prototype II could be steered with no rudder according to the dynamic hypothesis of my theory of dynamic interactions.

With the assumption of simultaneous noncoaxial rotations, the rigid body experiences nonhomogeneous velocity fields. These fields generate anisotropic acceleration fields, which can be interpreted as fields of inertial forces created in space through the effect of simultaneous noncoaxial rotations. The result of my experiments confirmed my initial assumption that a correlation can exist between rotational and orbital motions such that a body with intrinsic angular momentum [9,10] and translational velocity subjected to a real noncoinciding torque, transforms its dynamic action into a precession, which will generate a flat path with a closed orbit for a constant torque.

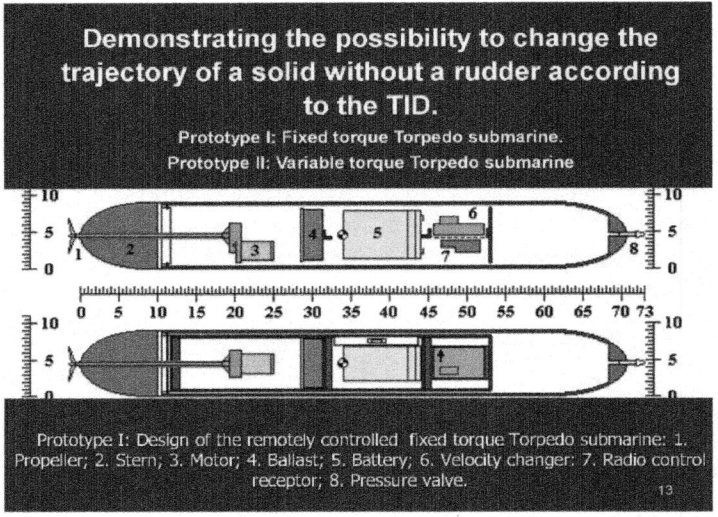

Figure 3: (Color online) Prototype II: design of remote-controlled torpedo submarine.

EQUATION OF MOTION

On the basis of the principle of conservation of motion [11], I obtain the equation of motion [12]. Consider an infinitely thin and flat disk with ratio d that experiences a torque M directed along the disk's symmetry axis (Zaxis). If this torque is instantaneous, it generates a constant angular velocity ω. A second torque M', noncoaxial with the former, will generate a new dynamic state, defined in classical mechanics as the gyroscopic effect and attributed to a supposed gyroscopic momentum D. Based on the principle of conservation of motion, we can interpret the gyroscopic momentum D as equivalent to the effect of the external torque M' and as being the torque that generates the second rotation, which is noncoaxial with the first. Therefore,

$$M' \equiv D \tag{2}$$

If the torque M' is constant in time, it will maintain its action on the body. Nevertheless, the gyroscopic momentum is quantified through multiple methods of classical mechanics by

$$D = I\Omega \times \omega \tag{3}$$

If we consider the disk with its own angular velocity ω and subject it to a new noncoaxial torque M', we find that it initiates a new angular velocity Ω about an axis perpendicular to the new torque M' and not about its own axis. Therefore, we infer that the field of inertial forces generated in the rotating frame by the new noncoaxial torque M' and acting upon a moving body with angular velocity ω and inertial momentum I about that axis of rotation (and thus with angular momentum L) will force the moving body to attain orbital motion at the angular velocity Ω defined by the scalar quotient

$$\Omega = M'/(I\omega) = M'/L \tag{4}$$

The angular velocity Ω can be observed at the same time as the initial angular velocity, ω which remains constant within the body. Therefore, instead of the discriminating Poinsot hypothesis [13-15], which supposes that angular momentums are coupled to each other and distinct from linear nonstatic momentums, I propose, for translational motion, the hypothesis that the translational velocity field couples to the anisotropic velocity field that is generated by the second noncoaxial torque, forcing the center of mass of the disk to modify its path with no external force applied. Therefore, we obtain an orbital motion Ω simultaneously with the constant intrinsic angular velocity ω of the body. This new orbital motion is generated by a noncoaxial torque and is defined by the rotation of the velocity vector, constant in module.

The new external torque M', which is supposed to act along the X-axis (Figure 4), will generate a rotation about the Z-axis so that if the initial translational vector V_0 were located on the XY plane, the resulting velocity v will remain on that plane after the rotation. This rotation may be described by the spatial rotation matrix $\overrightarrow{\Psi}$: and will, under my assumption, generate a rotation of the translational vector V_0 in the XY plane.

$$\begin{pmatrix} \cos\alpha & -\sin\alpha & 0 \\ \sin\alpha & \cos\alpha & 0 \\ 0 & 0 & 1 \end{pmatrix} \tag{5}$$

Given that no external force acts on the solid to modify its translational motion, its kinetic linear momentum must remain constant and therefore its translational velocity must remain constant. However, if we accept that the homogeneous translational velocity field couples with the tangential velocity field through the torque M', we can determine the body's new dynamic state. Under this hypothesis, the motion equation would be determined by the translational velocity of the body's center of mass, which has not varied in magnitude and therefore will be equal to the initial translational velocity of the body subjected to the spatial rotation mentioned above:

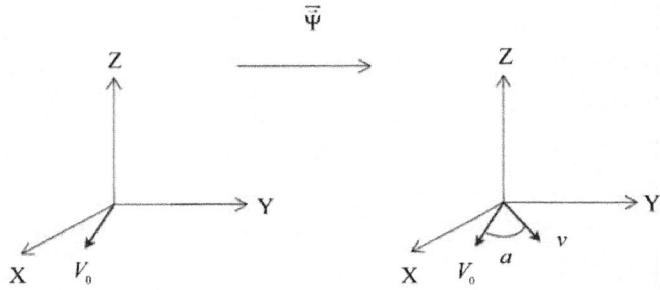

Figure 4: Coordinate system showing rotation by rotational operator $\vec{\vec{\Psi}}$, which transforms velocity V_0, through a single rotation α, into velocity v, both vectors are in the same plane (the XY plane in this example). That plane must also contain the acting torque M'.

$$v = \vec{\vec{\Psi}} \cdot V_0 \tag{6}$$

The nondiscriminating coupling proposed in my hypothesis is hence identified as a spatial rotation of velocity:

$$v = \begin{pmatrix} \cos\alpha & -\sin\alpha & 0 \\ \sin\alpha & \cos\alpha & 0 \\ 0 & 0 & 1 \end{pmatrix} \cdot V_0 \tag{7}$$

Thus, the equation of motion can be written as

$$v = \vec{\vec{\Psi}} \cdot V_0$$
$$= \begin{pmatrix} \cos(M')t/(I\omega) & -\sin(M')t/(I\omega) & 0 \\ \sin(M')t/(I\omega) & \cos(M')t/(I\omega) & 0 \\ 0 & 0 & 1 \end{pmatrix} V_0 \tag{8}$$

In a single rotation, the rotational operator $\vec{\vec{\Psi}}$ transforms the initial velocity V_0 into the velocity v, both situated in the same plane. We find that the rotational operator $\vec{\vec{\Psi}}$ is a function of sine or cosine of Ωt, which clearly indicates the relation between the angular velocity Ω of the orbit and the torque M' and the initial angular velocity ω. Thus, I have derived a simple mathematical relation between the angular velocity ω of the body and its translational velocity v.

Equation (6) is a general equation of motion for bodies with angular momentum that are subjected to successive noncoaxial torques. For this equation, the rotational operator $\dot{\Psi}$ serves as a matrix that transforms the initial velocity, by means of rotation, into the velocity that corresponds to each successive dynamic state.

In short, this simplified mathematical model implies that it would be possible that moving bodies subjected to successive noncoaxial torques would initiate orbital motion as a result of inertial dynamic interactions such that, while maintaining constant initial angular momentum and the second torque, its center of mass would follow a closed orbit without requiring any centripetal force (**Figure 5**).

Thus, we can associate dynamic effects to velocity and we find a clear mathematical correlation between rotation and translation. This mathematical connection allows us to identify a physical relation between transfers of rotational kinetic energy to translational kinetic energy and vice versa.

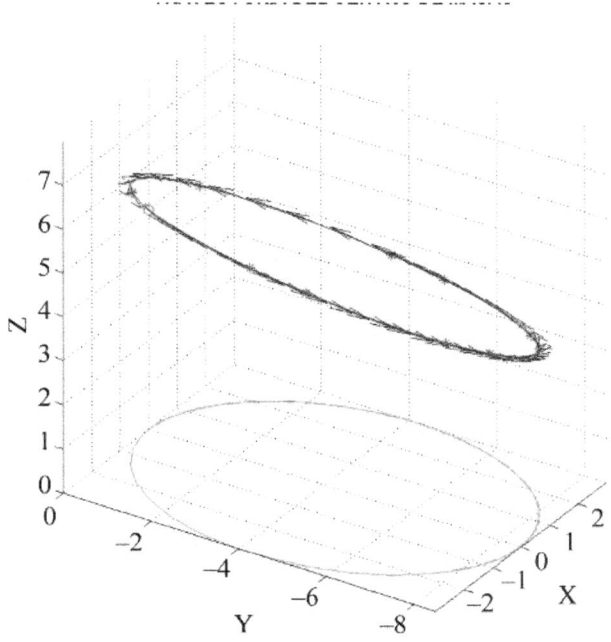

Figure 5: (Color online) Computer-simulated path of the center of mass of a moving object, which has intrinsic rotation and is simultaneously subjected to an external torque noncoaxial with its intrinsic angular momentum when both, the applied moment and translational velocity of the moving object are constant. For this simulation, the tangential velocity was 5 m/s.

A different approach is thus necessary to analyze the dynamics of bodies undergoing acceleration by rotation. In these cases, we cannot apply the same axioms and premises as used for inertial systems. In accordance with the proposed dynamic hypotheses, I simulated the threedimensional motion of an object in such a situation (i.e., with intrinsic angular momentum and simultaneously subjecting to an external noncoaxial torque) and obtained open or closed trajectories. The deduced equation of motion (Equation (6)) correctly predicts the path observed in my experimental model system (**Figure 5**).

CONCLUSIONS

The present text and the accompanying video Theory of Dynamic Interactions provide only a brief summary of the studies carried out over the last twenty years to propose a Rotational Dynamics of Interactions applicable to bodies subjected to multiple successive noncoaxial torques. The initial hypotheses are based on new criteria involving velocity coupling and have been confirmed by experiments and mathematical modeling that allows accurate physical simulations. In this study, I found a clear correlation between the initial speculations, original hypotheses, mathematical simulation, deduced physical laws, experiments, and mathematical models corresponding to the equations of motion that result from the proposed dynamics laws.

As a result of this dynamic investigation work we can propose the following conclusions:

1. There is a wide subject area not yet developed in rotational dynamics inasmuch as rigid bodies are subjected to accelerations caused by simultaneous non coaxial rotations.

2. This area of knowledge can be analyzed under relativistic and non relativistic mechanics. Hypotheses are based on new criteria about speed coupling and rotational inertia.

3. In the exposed experimental non relativistic tests carried out, we have concluded that new general laws of behavior can be obtained, based in the analysis of the dynamics fields created.

4. We have obtained an equation of motion for rigid bodies in translational motion with intrinsic angular momentum, when subjected to non-coaxial pairs, which defines the dynamic behavior of rigid bodies in these cases.

5. We find a clear mathematical correlation between rotation and translation. This mathematical connection allows us to identify a physical relation between transfers of rotational kinetic energy to translational kinetic energy and vice versa.

6.   The mathematical model implies that it would be possible that moving bodies subjected to successive noncoaxial torques would initiate orbital motion as a result of inertial dynamic interactions.

7.   While maintaining constant initial angular momentum and the second torque constant, the center of mass of the moving bodies would follow a closed orbit without requiring any centripetal force.

8.   The theory also allows to give an answer to an initial aporia: to be aware and to understand the physical and mathematical correlation between orbitation and intrinsic rotation.

An example of the theory is the feared roll coupling of the aeroplanes. It happens when a plane, which is flying a screw or any other kind of air acrobatics which implies, for example, a turn around its main inertia axe, starts a new steering manoeuvre with curved trajectory. According to the supported dynamic hypotheses, the non-homogeneous distribution of speeds, generated by the new non-coaxial rotation of the plane mass, couples to the field of translation speed, causing an unintentional deviation of the trajectory, as well as a possible loss of the plane control.

I developed this dynamic model and its background in two books about non-Newtonian mechanics: The Flight of the Boomerang [3], an essay in honor of the physicist Miguel Catalán, and A Rotating World [6].

The result of this project is the demonstration of a rational field theory that gives a new understanding of the behavior of matter. In my opinion, the application of these dynamic hypotheses to astrophysics, astronautics, and other fields of physics and technology will allow new, surprising, and stimulating advances.

The result of this project is the conception of an innovative dynamic theory, which specifically applies to rigid rotating physical systems and which has numerous and significant scientific and technological applications, especially in orbital dynamics, orbit determination, and orbit control. For instance:

•   Variation of the affecting torque, arises when subjecting intrinsic angular momentum bodies to new non-coaxial momentums.

•   To conceive an intrinsic rotating mobile solid, which could be exclusively controlled due to Dynamic Interactions.

•   To calculate the trajectory of any intrinsic angular momentum solid in space.

•   To propose a new steering system independent from a rudder or any other external element.

We can suggest advances in the studies and application related to orbital mechanics, guidance, navigation, and control of single or multi-spacecraft systems as well as space robotics and rockets [16].

For those who are interested in cooperating with this independent research project may request for additional information by writing me at gestor@ advanceddynamics.net or visiting www.advanceddynamics.net.

REFERENCES

1. P. Appell, "Traité de Mécanique Rationnelle," GauthierVillars, Paris, 1909

2. I. Newton, "Principia," Im. Du Chasteller, Paris, Proposition 2, 1757.

3. G. Barceló, "El Vuelo del Bumerán," Marcombo, Barcelona, 2006, p. 98.

4. M. E. Jouffret, "Théorie Élémentaire des Phénomènes que Présentent le Gyroscope, la Toupie et le Projectile Oblong," Berger-Levrault, Extract Revue d´Artillerie, París, 1874.

5. P. Gilbert, "Problème de la Rotation d'un Corps Solide Autour d'un Point," Annales de la Société Scientifique de Bruxelles, 1876, p. 316.

6. G. Barceló, "Un Mundo en Rotación," Marcombo, Barcelona, 2008, p. 208.

7. G. Bruhat, "Mécanique," Masson & Cie, Paris, 1955.

8. A. P. French, "Newtonian Mechanics (The M.I.T. Introductory Physics Series)," W. W. Norton & Company, New York, 1971.

9. L. D. Landau and E. M. Lifshitz, "Mechanic: Volume 1 (Course of Theoretical Physics)," 3rd Edition, Butterworth-Heinemann, Oxford, 1976,

10. L. D. Landau and E. M. Lifshitz, "Mecánica," Ed. S.A. Reverté, 1994, p. 24.

11. E. Mach, "Die Mechanik in Ihrer Entwicklung Historisch-Kritisch Dargestellt," Leipzig, Brockhaus, 1921.

12. H. Goldstein, "Classical Mechanics," Addison Wesley, Reading, 1994.

13. L. Poinsot, "Théorie Nouvelle de la Rotation des Corps," 1834.

14. Gilbert, "Problème de la Rotation d'un Corps Solide Autor d'un Point Solide," Annales de la Société Scientifique de Bruxelles, 1878, p. 258.

15. G. Barceló, "El Vuelo del Bumerán," Marcombo, Barcelona, 2006, p. 121.

16. G. Barceló, "On the Equivalence Principle," The 61st International Astronautical Congress, American Institute of Aeronautics and Astronautics, Prague, 27 September-1 October 2010.

Chapter 2

ROLE ON MOMENT OF INERTIA AND VORTEX DYNAMICS FOR A THIN ROTATING PLATE

Yoshihiro Kubota and Osamu Mochizuki
Faculty of Science and Engineering, Toyo University, Saitama, Japan

ABSTRACT

In this study, we focused on the lift generation with a thin rotating plate. The objective of this study is to understand the appropriate shape and the role of vortex for rotating thin plate. We determined the shape of the plate through free-flight tests of paper strips and investigated the aerodynamic characteristics of the rotating plate with the selected shape. The rectangular plate with an aspect ratio 7 was relevant from moment of inertia and bending stress. An endplate on a wing tip increased the stability on the lateral vortex structure behind the rotating plate. Velocity field measurement by Particle Image Velocimetry (PIV) showed that the lift force was generated twice in a rotating cycle.

INTRODUCTION

The aerodynamics of freely falling paper plate such as business cards was studied to understand for the fluttering or the tumbling. The purpose of this study is to understand the appropriate shape of plate, and the role of vortex to fly or glide the longer distance with a rotating thin plate. We are focusing on the aerodynamic characteristics of autorotation of thin paper plate. The autorotation is free to rotate with a fixed axis of a centroid of paper plate. The motion of paper plate during the autorotation is not only vertically but also horizontally. Moreover, the horizontal motion is not periodic oscillating motion like a swing motion. This means that the paper plate has the direction of a horizontal displacement. Mittal et al. showed that the paper plate during the autorotation generates the lift force [1]. So that if we controlled rotation of paper plate certainly to generate the enough lift force during the autorotation. This might become a small flier such a Micro Air Vehicle (MAV) [2]. This is a starting point of our study.

From the previous study, the aerodynamics of freely falling paper plate was known with the fluttering, the tumbling, and the autorotation. The experimental study on the frequency of tumbling plate was studied by Mahadevan et al. [3]. They concluded that the tumbling frequency was associated with the cross-sectional shape of plate as the chord length of paper and the thickness of paper. The increasing of frequency occurred with squared root of thickness. Hirata et al. reported the experimental study of an aerodynamic characteristics of tumbling plate [4]. They studied the relation among inertia moment, rotation, lift coefficient, and drag coefficient of paper as aerodynamic characteristics. Their results proposed the empirical formulation of aerodynamic characteristics from a function of Reynolds number.

From the numerical approach, Andersen et al. showed the numerical study of relation between the moment of inertia by the shape of plate and the transition between fluttering and tumbling [5]. Also, they reported both a comparison between numerical and an experimental studies. The experimental results indicated the transition between fluttering and tumbling was caused with the increasing thickness of paper. Pesavento et al. were studied the aerodynamic lift during a tumbling [6]. They were shown that an aerodynamic lift on a tumbling plate was dominated by the both falling velocity and angular velocity of stiff paper plate such as the business card. An influence of cross sectional shape for freely falling paper was studied numerically by Jin C. et al. [7]. The main difference with a cross sectional shape was angular velocity. For a tumbling, angular velocity dominates the lift force by Pesavento et al. [6], so that the shape for falling paper plate was needed to be considered for autorotation.

The main scope of our study is the shape of paper, and the stability during the autorotating plate. The influence of shape was focusing on the influence of inertia moment of paper, since the larger moment of inertia causes the difficulty of rotation. Since, the rotation of paper plate is key subject of autorotation. In addition, the starting point of this study is to understand the generation of lift force with an autorotation for a small flier, thus the stability of autorotation is necessary to studied. The influence of endplates of rotating wing was studied for the improvement of aerodynamic stability.

We determined the aspect ratio of the plate by using a falling paper and observed that a plate with an aspect ratio of 7 achieved the longest flight distance, which means it must have the highest lift/drag ratio. Next, we tested the stability performance with a thin auto rotating plate with an aspect ratio of 7. We employed particle image velocimetry (PIV) to the investigate vortex formation around the rotating plate. In addition, we discuss herein the relation between the flight performance and the shape of wing.

EXPERIMENTAL APPARATUS AND METHOD

We performed three different experiments: 1) free-flight tests with paper strips of varying shapes and a test model with an autorotating thin plate; 2) flow visualization near the tip of a rotating plate; 3) PIV measurements of vortices produced by the rotating plate.

To select the shape of the rotating plate that will be used in the test model, we checked the free-flight performance of paper strips with different shapes, as shown in **Figure 1**. The strips shown in**Figure 1**(a) have the same aspect ratio, which is defined as the ratio of the spanwise length to the chord length at the tip of the plate. The strips shown in **Figure 1**(b) have the same area but different aspect ratio and are all made of paper.

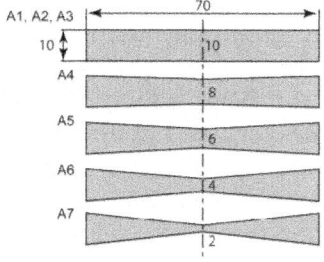

(a)

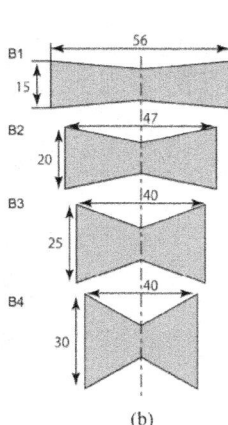

(b)

Figure 1: Different shapes of paper strips. (a) Identical aspect ratio, but different chord length at the center position; (b) Identical area, but different aspect ratio. Weights of A1, A2, and A3 are 0.038, 0.049, and 0.064 g, respectively. Dimensions are in mm.

The test model shown in Figure 2, which had a thin plate rotating about its spanwise axis, was used to observe flight performances. The size of the plate, which served as a rotating wing, was 20 mm along the chord and 70 mm in the spanwise direction. Thus, the total aspect ratio of the two plates aligned on each side of the center sphere was 7. The center spherical body made from styrene foam was 30 mm in diameter. The total weight of the model was 30 g. Future plans include transforming the spherical body into a fuselage in which a motor, actuators, sensors, and a tiny camera are packed. The photograph in Figure 2(b) shows the test model with end plates affixed at both tips of the plate.

An experimenter threw the test model with paper strips in the forward direction by hand, releasing it 2 m above the floor. Once released, the model fell freely. A 0.25×0.25 m^2 square grid was drawn in an area of 5×5 m^2 on the floor to record the landing point of the test model. The averaged falling speed was 1.3 m/s. Thus, based on the speed and chord length, the Reynolds number was 1.7×10^3. The spin parameter, which we define later, was 0.4. To check the effects of the end plates (attached at the plate tips) on flight performance, the test model with end plates (as seen in Figure 2) was flight-tested under conditions similar to those for the model without the end plates. Five hundred flight tests were conducted for one model in a room with no flow (i.e., no air conditioner). The experimenter started the flight tests after practicing the model throw so as to minimize the influence of the launch on the experimental results.

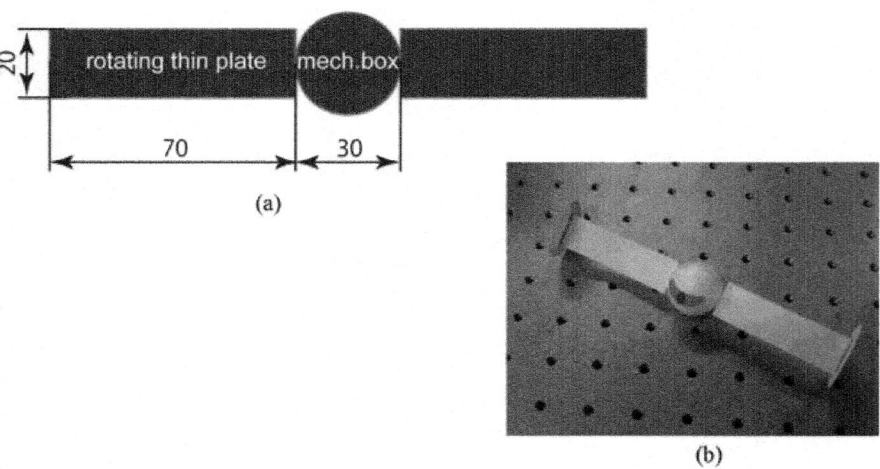

(a)

(b)

Figure 2: Test model. (a) Schematic with dimensions in mm; (b) photograph of test model with end plates on left and right extremities.

We investigated the three-dimensional effects of the finite plate by visualizing the fluid flow using a dye in a circulating open water channel as shown in Figure 3. The plate immersed in the water had a chord length of 40 mm and spanwise length of 180 mm. Revolutions of the plate were controlled by a motor to change the peripheral speed of the plate relative to the main flow velocity. The ratio of the peripheral speed of the plate to the main flow velocity is called the spin parameter S and is defined as follows:

$$S = \pi Nc/(60U)$$

(1)

Here N represents the revolutions per minute of the rotating plate, c is the plate chord length, and U is the main stream velocity. The main stream velocity was 0.1 m/s in the water channel experiment. Based on the chord length and main stream velocity, the Reynolds number was 4×10^3. We visualized the flow near the wing tip by using a dye injected upstream of the rotating wing and recorded streak-line interactions by a high-speed camera. Vortex formation around the rotating wing and velocity fields were observed by PIV. Particles 100 mm in diameter were used to detect velocity fields.

RESULTS AND DISCUSSION

Free-Flight Test

Figure 4 shows the relationship between the flying time and moment of inertia about the spanwise axis of the paper strips shown in Figure 1. The paper strips shown in Figure 1(b) have the same area but different moments of inertia. In this case, the strip with a smaller moment of inertia flies longer, as seen in Figure 4. Since the strip with the smaller moment of inertia rotates more easily, this result indicates that the rotation speed is responsible for long flight times.

On the other hand, the strips shown in Figure 1(a) have the same aspect ratio but different chord lengths at the center position. Although the moment of inertia of strip A7 is small, its flight time is short. Furthermore, the decreasing of chord length at the center caused the bending of wing easily from the observation. Since, the maximum bending stress increased with the decreasing of chord length at the center. As a result, the rectangular paper has the longest flight time. Therefore, the suitable shape of plate for autorotation is the rectangular as the flier. For the model test, the rectangular wing was used.

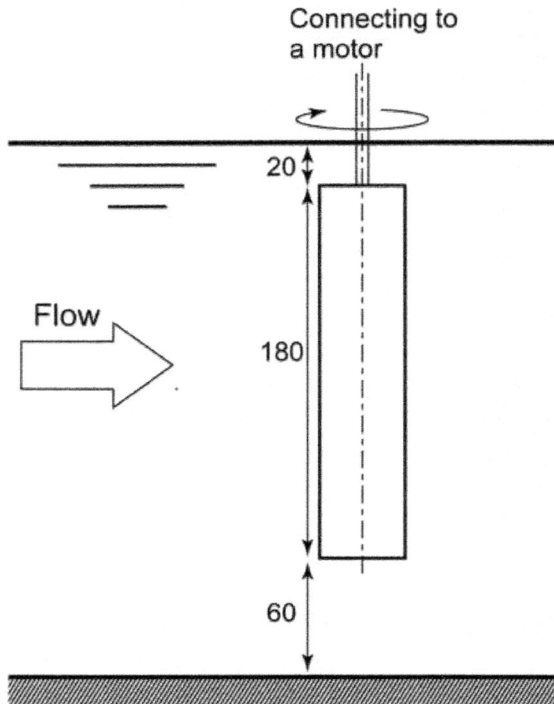

Figure 3: Experimental setup for dye visualization and water-tunnel PIV measurements. Dimensions are in mm.

Figure 5 shows the distribution in landing position for the model plane without end plates that is shown in Figure 2. The flight locus of the model plane is always spiral in this case, as shown in the photograph in Figure 5. The model plane tilts just after being launched, and this tilt angle is unpredictable, making it difficult to predict the landing position. Thus, the landing positions are distributed around the circumference of a circular strip, as seen in Figure 5. The maximum distance between the launch and landing position is approximately 2.7 m. We consider the flight to be strongly influenced by small disturbances caused by a tip vortex.

Then, we attached end plates at both tips of the plate, as shown in the photograph in Figure 2. The end plate of a fixed wing reduces the induced drag by preventing interaction between the flow along the upper and lower surfaces of the wing. The landing positions of the model plane with end plates are converged into a small area in front of the launch position, as shown in Figure 6. Thus, the model plane with end plates flies straight forward, as seen in the photograph in Figure 6.

Since the experimental conditions (except for the end plates and throwing manner) are similar to those in the previous experiment, the difference in flight paths is attributed to the effect of the end plates. This shows that the end plate is effective in stabilizing the flight of the model plane. Since the pressure distribution on the wing surface is influenced by lateral vortices generated from the edge of the rotating plate, the interactions between the tip of the vortex and lateral vortex affect aerodynamic forces, which in turn affect the flight of the model plane. Thus, the end plates cut the cross talk between the upper and lower surfaces of the wings. In fact, end plates are well known to reduce the induced drag. The use of end plates avoids the uncertainty caused by the three-dimensional deformation of the connected vortex at the wing edge, which facilitates stable flight.

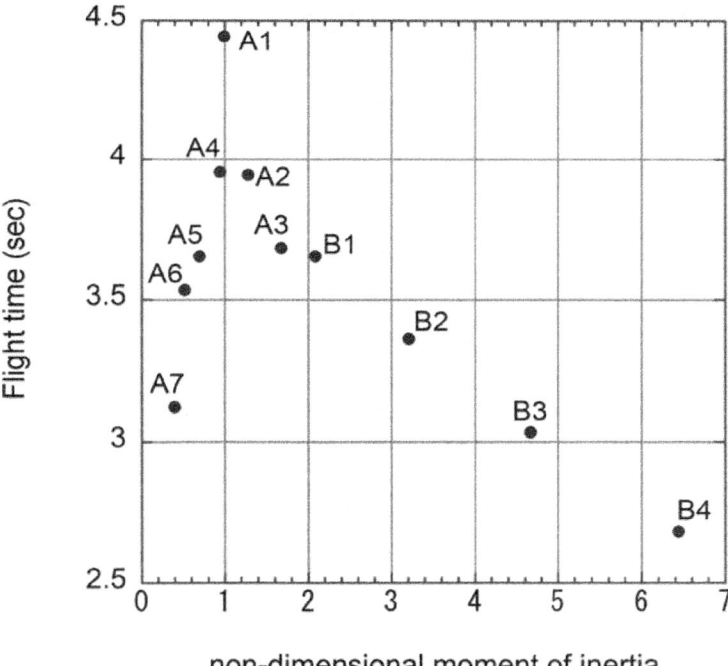

Figure 4: Flying time plotted versus moment of inertia of paper strips. Each label corresponds to those shown in Figure 1.

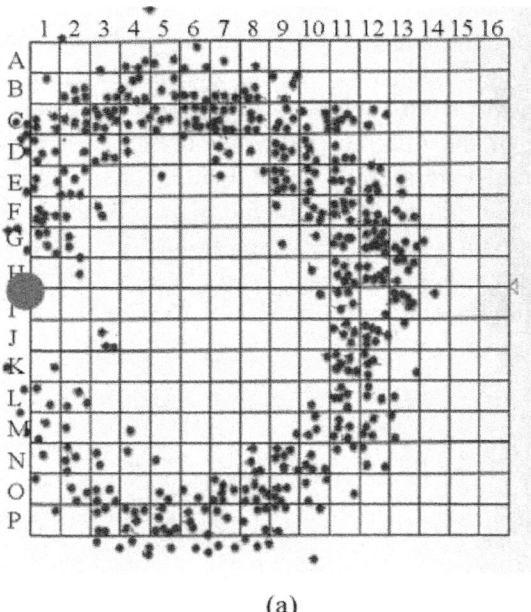

(a)

(b)

Figure 5. Distribution in the landing position (a) and typical flight (b) path for model plane without end plates.

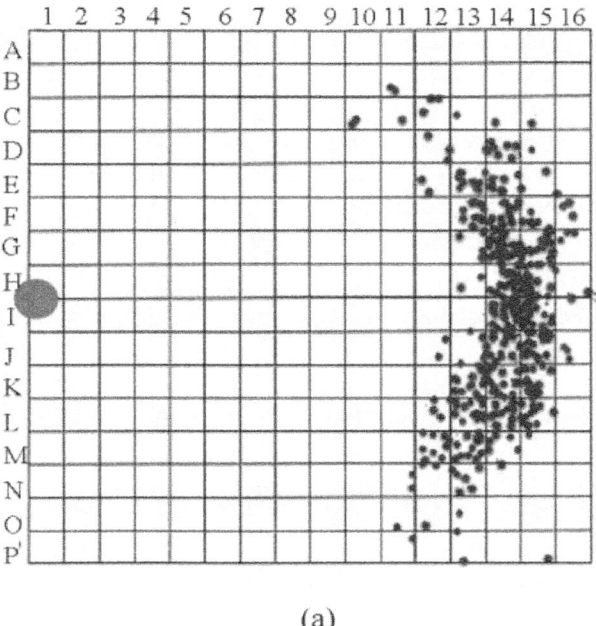

(a)

(b)

Figure 6: Distribution in the landing position (a) and typical flight path (b) for model plane with end plates.

Flow Visualization near Plate Tip

To observe the effects of the end plate, flows near the tip of the main plate were observed using a dye. Figure 7 compares the streak lines near the main plate tip for a plate with and without an end plate. The streak lines shown in Figure 7(a) (without end plate) are entangled behind the plate, which indicates that the flow behind a rotating plate is disturbed by the trailing vortex near the tip. A lateral vortex separated from the forward edge of the rotating plate is easily affected by this disturbance.

The distribution of the surface pressure induced by the separated vortices must be different at both tips. Thus, the lift forces on both tips are not balanced; thus, the model plane tilts if there are no end plates. In contrast, the streak lines near the tip of the plate with the end plate are laminar, as seen in Figure 7(b). This shows that the end plate is useful for stabilizing the lateral vortex structure behind the rotating plate. Thus, the model plane with the end plates has balanced lift forces, and thus, it flies straight, as seen in the photograph in Figure 6.

PIV Measurements near Rotating Plate

With PIV, we observed lateral vortices at the leading and trailing edges of the rotating plate. To investigate the temporal evolution of lateral vortices, we captured the flow field at phase steps of $\pi/4$, which was accomplished with a high-speed camera and a double-pulse-laser-light sheet that detected the flow field during the rotation of the plate in an open water channel. The flow fields were averaged at each phase. The temporal evolution of the flow field in the middle section of the rotating plate (without the end plate) is shown in Figure 8 for S = 0.6 and 1.0.

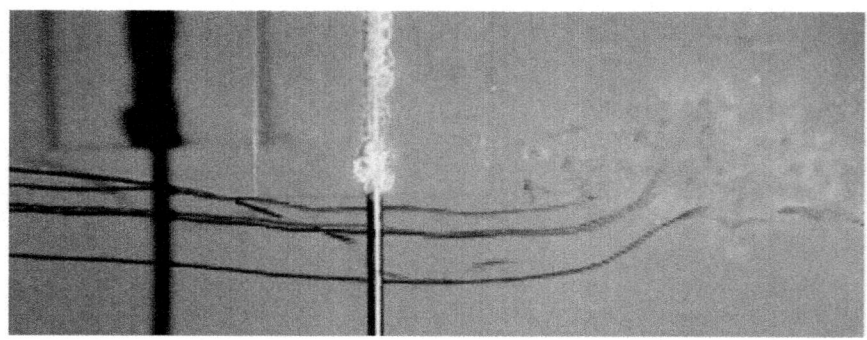

(a)

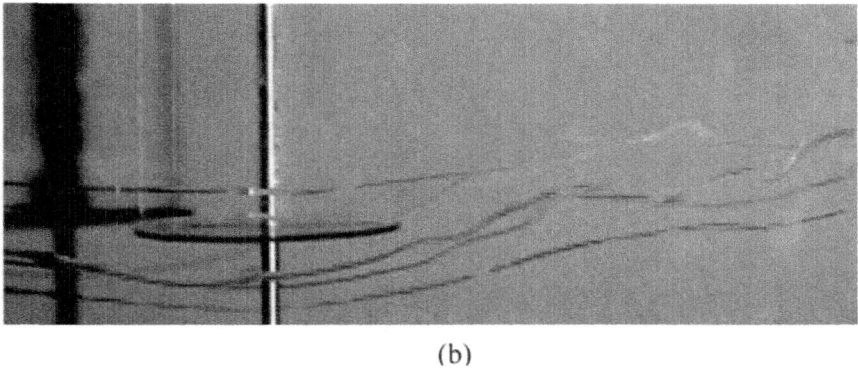

(b)

Figure 7: Streak lines near the tip of a rotating plate (a) without an end plate and (b) with an end plate.

The velocity vectors and stream lines are visible in Figure 8 and show the formation of a lateral vortex behind the rotating plate. The vectors are color coded according to speed. Time progresses from the left panel to the right. The time interval between each figure corresponds to a $\pi/4$ phase step in one revolution cycle. The labels A and B at the plate edges identify the edges during a revolution. The edge moves backward, and the relative velocity of the edge with respect to the main flow velocity is small when $S < 1$. The B edge moves forward, and the relative velocity of the edge with respect to the main flow velocity is always large. The changes in circulation generated at the A and B edges are estimated over an angular range $0 < \omega t < \pi$ by the following relationships:

$$d\Gamma_A/dt = (1/2)\{U - v\sin(\omega t)\} \text{ on A-side edge}$$

(2)

$$d\Gamma_B/dt = (1/2)\{U + v\sin(\omega t)\} \text{ on B-side edge}$$

(3)

Here ω is an angular speed, and v defined by $v = \omega t/2$ is the linear speed at the edge of the rotating plate. The term $U \pm v \sin(\omega t)$ represents the relative flow speed at each edge. From $\omega t = 0$ to π, the change in the circulation of vortices shed from the A edge is smaller than that from the B edge. The change in the flow field during half a rotation is also shown in Figure 8 (the flow field is the same for the other half cycle). The flow field at $S = 0.6$ was obtained in a manner similar to that previously discussed for $S = 1$. The strong lateral vortex seen in the top-left panel of Figure 8 is shed from the B edge during the previous half turn. The strong vortex is located below the B edge and rotates counterclockwise (ccw). This vortex is regarded as a starting vortex of a wing that moves suddenly. As a result, the rotation of the plate deflects the mean flow downward, which generates lift.

In contrast to the strong vortex, the weak lateral vortex is shed from the A edge and rotates in clockwise (cw). The weak vortex seen in the top-left panel of Figure 8 is shed from the A edge during the previous half turn and is located near the B edge at $\omega t = 0$. Since the small vortex travels near the upper surface of the plate until it is shed from the A edge, we speculates that it affects the pressure on the upper surface of the plate while the plate rotates. The shedding frequency of these vortices is two times the rotation frequency. Mochizuki et al. (1987) analyzed aerodynamic characteristics with an impulsively rotated plate. They showed that the generation of lift force was caused twice in a rotating cycle with a plate rotated on a center of axis. Thus, the lift force is generated two times per rotation.

The change in flow patterns at $S = 1.0$ is shown in the bottom row of Figure 8. By comparing figures for $S = 1.0$ at a given phase with those for $S = 0.6$ at the same phase, we obtain the effects of the rotation speed on vortex formation. For $S = 1.0$, the peripheral velocity of the backward-moving edge (i.e., the A edge) of the plate is equal to the main stream velocity. Thus, the relative velocity between the main stream and the A edge is zero when the wing angle is $90°$. According to Equation (2), the circulation change is static at that time; thus, the supply of vorticity to the lateral vortex shed from the A edge is smaller than that in the case of $S = 0.6$. Thus, the shear layer is hard to roll up on side A. In contrast to this vortex, the vortex shed from the B edge is stronger than that in the case of $S = 0.6$ because of the higher relative velocity. The strong vortex forms slightly below the edge. This affects the surface pressure induced by the strong vortex. This makes the prediction of aerodynamic forces difficult, because the induced pressure results from mutual effects involving both the strength of the vortex and the distance from the vortex to the surface.

CONCLUSION

The influence of the shape of plate and the role of vortex dynamics were investigated, herein we propose using a rotating plate about its spanwise axis. We conducted numerous flight tests using a model plane to understand the factors influencing the stability of its flight. The rectangular plate with an aspect ratio of 7 was the suitable for the flight with an auto rotating. The difficulty of rotation was examined with an increasing of moment of inertia. In addition, to understand three-dimensional effects and the unsteadiness of the tip vortex, we observed the formation of vortices near the tip of the rotating plate by a real-time PIV system. The lateral vortex that separates from the forward and backward edges is found to be contaminated by the tip vortex. However, attaching end plates at the tips effectively prevents this contamination, and we demonstrated stable flight of the model plane with end plates attached. Vortices generated from the backward-rotating edge are weaker than those generated from the forward-rotating edge. However, the former move near the wing surface together with the plate; therefore, the induced surface pressure is larger.

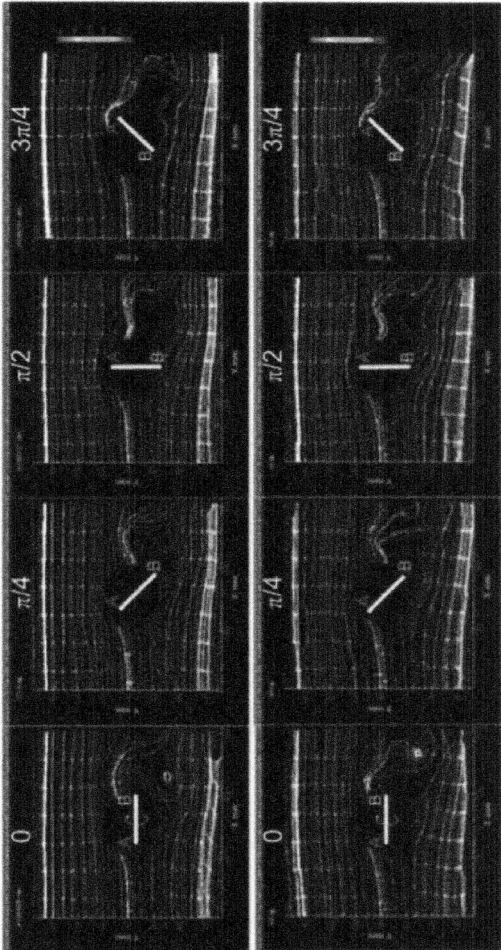

Figure 8: Time evolution of the flow field around the rotating wing without end plates in an open water channel. Velocity was measured by PIV. Lines represent stream lines. Color of a vector represents the magnitude of velocity. The passage of time is shown from left to right figures. Time interval of each figure is π/4 phase of a revolution cycle. Symbols A and B at edges of the plate distinguish the edges during the revolution in one cycle. (a) S = 0.6 and (b) S = 1.0.

REFERENCES

1. R. Mittal, V. Seshadri and H. S. Udaykumar, "Flutter, Tumble and Vortex Induced Autorotation," Theoretical and Computational Fluid Dynamics, Vol. 17, No. 3, 2004, pp. 1-6. doi:10.1007/s00162-003-0101-5

2. S. A. Ansari, R. Zbikowski and K. Knowles, "Aerodynamic Modeling of Insect-Like Flapping Flight for Micro Air Vehicles," Progress in Aerospace Science, Vol. 42, No. 2, 2006, pp. 129-172. doi:10.1016/j.paerosci.2006.07.001

3. L. Mahadevan, W. S. Ryu and A. D. T. Samuel, "Tumbling Cards," Physic of Fluid, Vol. 11, No. 1, 1999, pp. 1-3. doi:10.1063/1.869919

4. K. Hirata, M. Hayakawa and J. Funaki, "On Tumbling of a Two-Dimensional Plate under Free Flight," Journal of Fluid Science and Technology, Vol. 6, No. 2, 2011, pp. 177-191.doi:10.1299/jfst.6.177

5. A. Andersen, U. Pesavento and Z. J. Wang, "Analysis of Transitions between Fluttering, Tumbling and Steady Descent of Falling Card," Journal of Fluid Mechanics, Vol. 542, 2005, pp. 65-90. doi:10.1017/S002211200500594X

6. U. Pesavento and Z. J. Wang, "Falling Paper: NavierStokes Solutions, Model of Fluid Forces, and Center of Mass Elevation," Physical Review Letters, Vol. 93, No.14, 2004, Article ID: 144501. doi:10.1103/PhysRevLett.93.144501

7. C. Jin and K. Xu, "Numerical Study of the Unsteady Aerodynamics of Freely Falling Plates," Communications in Computational Physics, Vol. 3, No. 4, 2008, pp. 834- 851.

Chapter 3

VISCOSITY TRANSIENT PHENOMENON DURING DROP IMPACT TESTING AND ITS SIMPLE DYNAMICS MODEL

Chen Liu, Yoshikazu Tanaka, and Yukio Fujimoto
Department of Transportation and Environmental Engineering, Graduate School of Engineering Science, Hiroshima University, Higashi-Hiroshima, Japan

ABSTRACT

Most soft materials behave as if they were hardened when subjected to an impact force. The strain rate dependence of viscosity resistance is the reason for this behavior. The authors carried out drop impact tests on several types of soft materials under the condition of a flat frontal impact. The impact force waveform of soft materials was found to consist of a thorn-shaped waveform and a succeeding mountain-shaped waveform. Based on our experimental observations, we believe that a large viscosity resistance is rapidly changed to a small resistance in the course of the impact. In the present study, the cause of this distinct waveform is discussed based on a dynamics model. The study applies a standard linear solid (SLS) model in which the viscosity transient phenomenon is considered is applied. Three types of impact force waveforms of actual soft materials are simulated using the SLS model. Some features of the impact force waveform of soft materials can be explained using the SLS model.

INTRODUCTION

Most soft materials have the properties of both viscosity and elasticity. Elastic properties are only slightly affected by the strain rate dependence. On the other hand, viscosity properties (including plastic deformation) are strongly affected by the strain rate dependence. A soft material with a strong viscosity resistance behaves as if it were hardened under a fast load, such as an impact load. For example, in a static state, agar jelly is softer than rubber. However, its impact force waveform has a shorter impact period than that of rubber.

In order to observe the features of impact force waveforms of soft materials, we conducted drop impact experiments for several types of soft materials [1] . The examined materials are agar jelly, sponge, sponge rubber, nitrile rubber, oil clay, paper clay, low-rebound urethane foam, ham, eligible meat, konjac, and cork. The experiments were carried out under a flat frontal impact condition.

The experiments reveal that the impact force waveform of a soft material consists of a thorn-shaped waveform, followed by a mountain-shaped waveform. The thorn-shaped waveform is a spike-like waveform, which is observed in the rising segment of the waveform. This type of thorn-shaped waveform appears in all the results for all of the soft materials except cork sheet. The existence of the thorn-shaped (spike-like) waveform is also reported in a previous study based on a Kolsky bar experiment [2] .

From the experimental observations, we believe that the thorn-shaped waveform is caused primarily by the viscosity resistance, and the mountain-shaped waveform is caused primarily by the elastic resistance. In other words, a transient occurs from the viscosity-dominant waveform to the elasticity-dominant waveform during the impact period.

In the present study, we attempt to simulate the impact force waveforms of soft materials by applying a dynamics model considering the viscosity transient mechanism. The standard linear solid (SLS) model is used in the simulation [3] [4] . The soft material is modeled as a combination of a spring and a dashpot. The viscosity transient is defined by the rapid change in viscosity resistance of the dashpot. The simulated waveforms are compared with the actual waveforms measured in the experiment.

IMPACT FORCE WAVEFORM OF A SOFT MATERIAL

Figure 1(a) shows a soft material subjected to a flat frontal impact. The impact object is a free drop hammer. Figure 1(b) shows typical thorn-shaped waveforms of the impact force. The features of the thorn-shaped waveform include a steep slope θ on the rising segment and a thorn-shaped waveform similar to the yield point of low-carbon steel. After that, at a certain point, the slope α becomes gentler and generates a second mountain- shaped waveform. The rising segment of the thorn is usually similar to a straight line.

Figure 1(c) summarizes the occurrence mechanism of thorn-shaped waveforms obtained by the experimental observation. The impact force waveform of soft materials is a mixture of two types of waveforms. The first (thorn-shape) waveform exhibits a large viscosity resistance (red curve), and the second (mountain-shape) waveform exhibits a small viscosity resistance (dashed black curve). This resistance change occurs suddenly at the thorn peak. In other words, during the impact period, the phenomenon changes from a viscosity-dominant waveform to anelasticity-dominant waveform. The compressive stress at the thorn peak is considered as the trigger of this transient. The soft material surface is struck twice by the relative movement of collision objects. The reason is that the slope α becomes smaller than the initial slope θ because the viscosity discontinuity is induced by the first strike.

CONDITION OF FLAT FRONTAL IMPACT AND COMPACT DROP TEST EQUIPMENT

Through a number of experiments, we found that the thorn-shaped waveform could be clearly measured under the condition of a flat frontal impact [1] [5] . In Figure 2, the impact object is solid and has a flat contact surface. The soft material that receives the impact force is a plate-like material with a uniform thickness without surface irregularities. The impact object strikes the soft material in the normal direction. The thorn-shaped waveform can be steadily measured under the above condition. When the bottom surface of the impact object is curved (Figure 2(b)), when the impact object is tilted (Figure 2(c)), or when the upper surface of the soft material is curved (Figure 2(d)), the thorn-shaped waveform becomes unclear. Here, "unclear" does not mean that thorn- shaped waveform does not occur. Rather, the thorn-shaped waveform is not obvious in the contact area that expands gradually from the local area, and the phenomenon is diluted.

Figure 3 shows the free drop test equipment used in this study. The device is braced by vertical columns on the left and right of the bottom plate. The top of the side column is connected with a metal plate. The drop hammer is made of an aluminum circular disk plate and a stainless cylinder (mass of 1.9 kg). The contact surface of the hammer bottom is a disk plate with a diameter of D = 60 mm. The upper sensor is an orbicular pad sensor and is attached near the drop hammer bottom. The lower sensor is a pad sensor of 100 mm square and is attached to the frame bottom. Both sensors are made from piezoelectric film. A high-speed voltage recorder with a sampling rate of 20 - 200 kHz is used for recording the sensor output. The output of the upper sensor is modified by 8% to account for the mass of the disk plate.

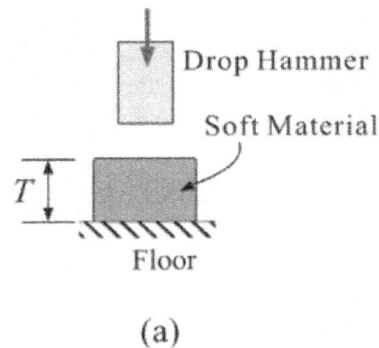

(a)

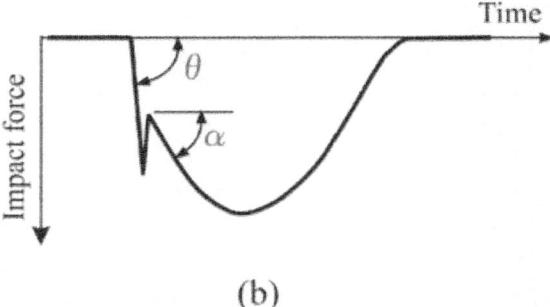

(b)

(Q): Thorn-shape waveform
(R): Mountain shape waveform

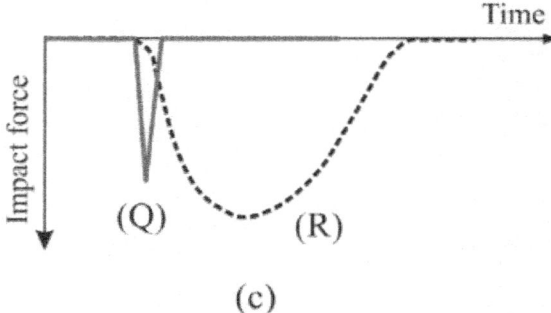

(c)

Figure 1: Impact force waveform of a soft material. (a) Flat frontal impact by a free drop hammer for a plate-like soft material; (b) Typical impact force waveform of a soft material including a thorn-shaped waveform; (c) Two types of waveforms contained in the impact force waveform of a soft material.

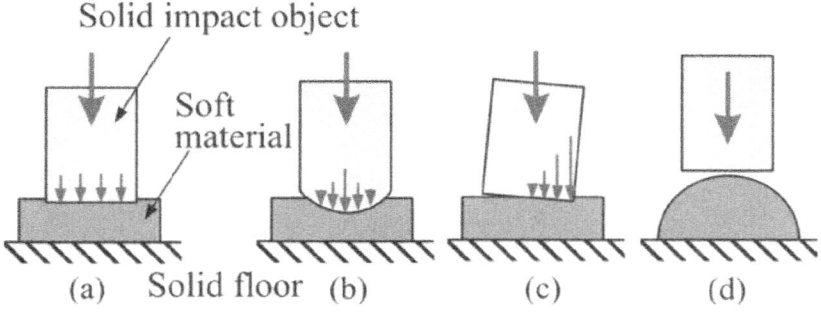

Figure 2: (a) Flat frontal impact condition for which the thorn appears clearly; (b), (c), and (d) Flat frontal impact condition for which the thorn is indistinct.

(a)

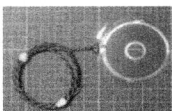

Upper sensor

Lower sensor

(b)

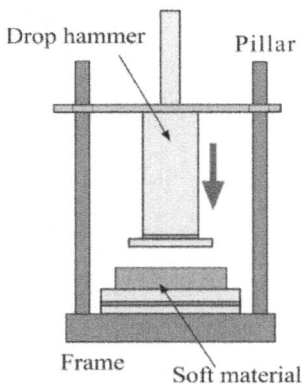

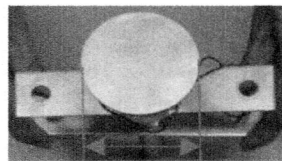

<u>Impact area</u>

Figure 3: Drop test equipment and the upper and lower sensors used in the device.

Based on rigorous tests using a servo-controlled testing machine and a high-rigidity load cell, it was confirmed that both sensors could measure the impact force very accurately when the contact time was in the range between 0.1 msec and 2 sec. The drop height h is the height from the upper surface of the soft material to the bottom surface of the drop hammer. The impact velocity is V_0 = 1.4 m/sec when h = 0.1 m, V_0 = 2.73 m/sec when h = 0.38 m, and V_0 = 3.13 m/sec when h = 0.5 m.

EXAMPLES OF MEASURED IMPACT FORCE WAVEFORMS

Figure 4(a) shows the impact force waveforms of a gel sheet (product name: a gel) with a T = 20 mm thickness. Figure 4(b) is an enlarged view of the thorn part. The drop height is h = 380 mm ($\dot{\varepsilon}_0$ = 136 sec^{-1}). The thorn- shape waveform can be seen clearly in both the upper and lower sensor outputs. In the enlarged view, we can observe that the slope θ of the rising segment is larger than the slope α of the segment past the thorn. The impulse of the thorn part is quite small compared with the succeeding mountain-shape waveform.

Figure 5(a) is the impact force of a nitrile rubber (NBR) plate with a T = 30 mm thickness. The drop height is h = 380 mm ($\dot{\varepsilon}_0$ = 91 sec^{-1}). The upper sensor output shows a relatively large thorn, and the lower sensor out- put shows a smaller thorn. The slopes θ of the upper and lower sensor outputs are almost the same. Figure 5(b) compares the impact force waveforms of the upper sensor for drop heights of h = 100, 200, and 380 mm. In the figure, when the impact velocity V_0 is faster, the slope θ of the rising segment and the height of the thorn peak are larger. This means that the slope θ and the height of the thorn peak are dependent on the impact velocity or strain rate.

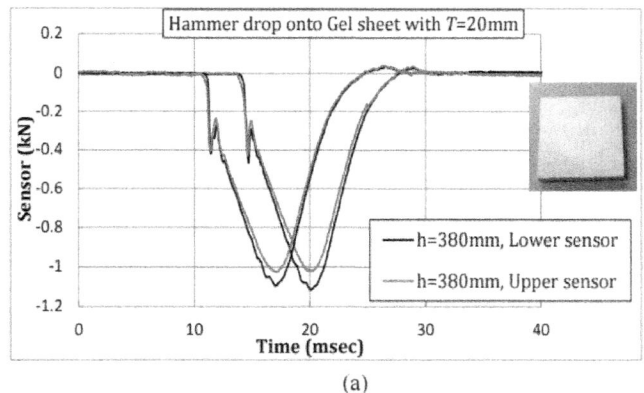

(a)

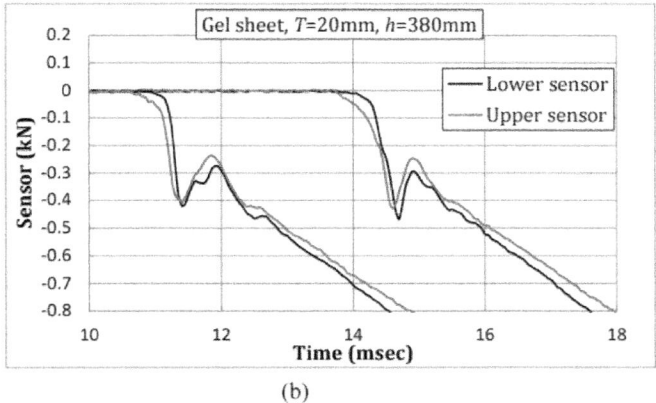

(b)

Figure 4: Impact force waveform induced by the hammer drop onto gel sheet with 20 mm thickness.

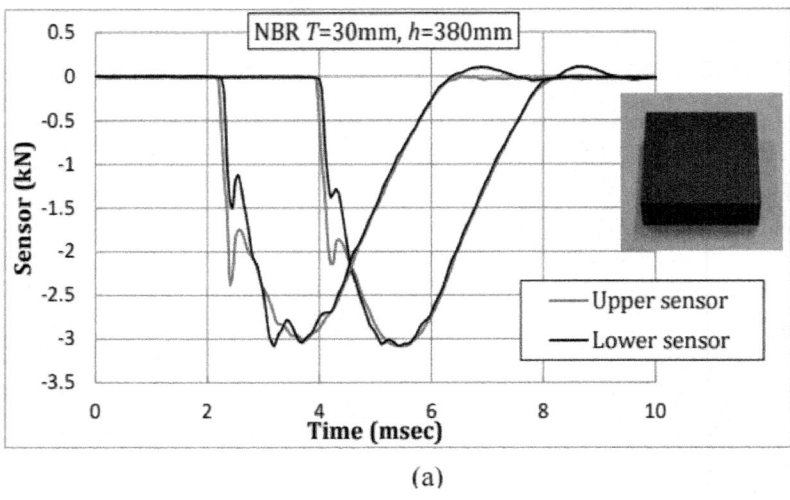

(a)

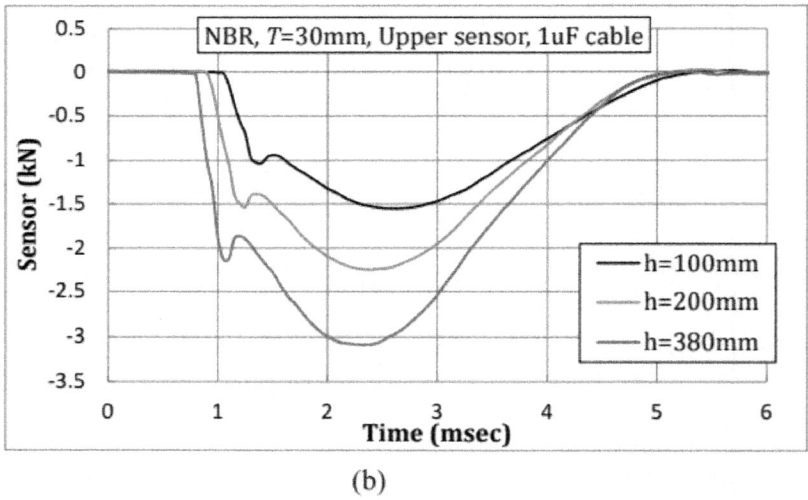

(b)

Figure 5. Impact force induced by the hammer drop onto NBR with T = 30 mm.

Figure 6 shows the impact force waveform of an oil clay plate with a T = 50 mm thickness. The drop height is h = 100 and 530 mm. A large thorn is seen by the upper sensor, and a small thorn is seen by the lower sensor. As the drop height increases (V_0 becomes faster), the slope θ becomes larger, and the thorn peak becomes taller. The thorns are more obvious than those of other materials. From the enlarged view, the thorn shape is close to an isosceles triangle.

VISCOSITY TRANSIENT PHENOMENON OF A SOFT MA-TERIAL UNDER THE DROP TEST

In the drop impact experiment, strain rate varies with time. The strain rate is large at the beginning of the collision. The strain rate is reduced with the compression of the material and becomes zero at the time of maximum compression.

Therefore, the strain rate dependency of the material changes at every moment during the impact. If it is assumed that the viscosity resistance decreases gradually along with the decrease of the strain rate, the thorn- shaped waveform will not be generated. From the experimental observations, however, we believe that the viscosity resistance changes rapidly or discontinuously (viscosity transient) in the course of the collision process.

Figure 7 shows a schematic diagram of the relationship between the viscosity resistance and the strain rate. We believe that two phases of viscosity resistance appear in the impact force results for soft materials. One phase is an excessively large viscosity resistance (red curve), which appears at the first stage, and the other phase is a small viscosity resistance (black dashed curve), which appears in the second stage. This viscosity transient (green dashed arrow) occurs at the peak of the thorn.

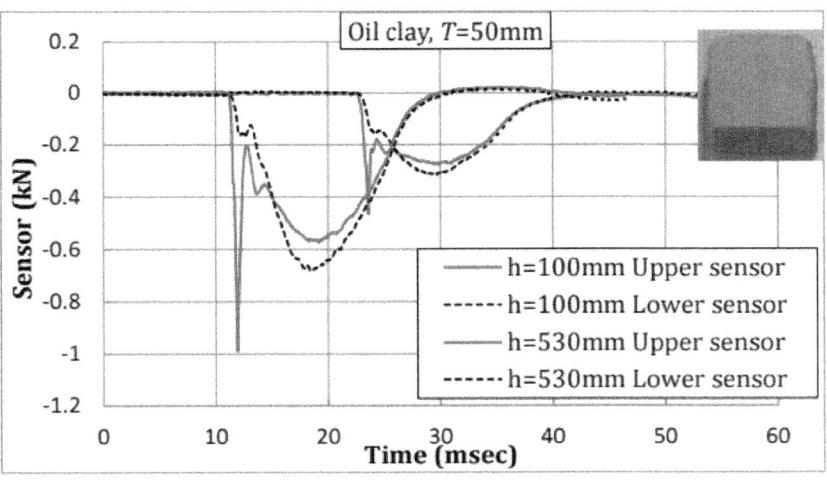

Figure 6: Impact force induced by the hammer drop onto oil clay with 50 mm thickness.

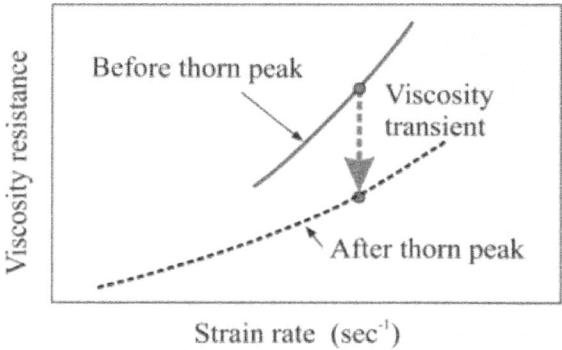

Figure 7: Viscosity resistance vs. strain rate relationship.

In the case of some colloidal suspensions, the viscosity changes rapidly at a certain shear rate. This phenomenon is known as the viscosity discontinuity [6] . This discontinuity also occurs under a reversible load [7] . Based on the experimental observations, we believe that this viscosity discontinuity phenomenon also occurs in most soft materials (soft solids). The impact force waveform of the soft material takes on a distinctive shape as a result of the viscosity discontinuity. In the next chapter, the impact force waveform is simulated considering the viscosity transient.

SIMULATION OF THE IMPACT FORCE WAVEFORM BY THE VOIGT MODEL AND THE STANDARD LINEAR SOLID (SLS) MODEL

Figure 8 shows the condition under which the drop hammer (mass: m = 1.9 kg) collides with the plate-like soft material at a speed V_0. The thickness of the soft material is T. The soft material is modeled by the Voigt model or the SLS model. The mass (drop hammer) is not connected to the spring and dashpot. The Voigt model can be represented by a viscous damper c and an elastic spring k_1 connected in parallel, as shown in the figure. The SLS model adds an elastic spring k_1 to the Maxwell model (c, k_2) in parallel. The impact velocity V_0 = 2.73 m/s, which corresponds to the drop height h = 0.38 m of the hammer, is given to the mass as the initial condition. The differential equation of the SLS model is given as Equation (1), where x_2 is the displacement of the hammer position, and x_1 is the displacement of the dashpot. The initial strain rate is $\dot{\varepsilon}_0 = V_0/T$ (s^{-1}). The Runge-Kutta method is used for solving the equation of motion.

$$\begin{cases} m\ddot{x}_2 + k_1 x_2 + k_2\left(x_2 - x_1\right) = 0 \\ k_2\left(x_2 - x_1\right) = c\dot{x}_1 \end{cases}$$

(1)

First, an analysis is carried out using the Voigt model. The spring coefficient k_1 is assumed to be $k_1 = 1000$ kN/m, which is constant throughout the collision. As mentioned above, we believe that the viscosity transient occurs in the early stage of the impact force. This influence is taken into consideration through the damping coefficient c of the dashpot. Figure 9 shows the assumed c as a time-dependent value. Here, c is c = 500 kg/s (0 < t < 0.3 ms). Then, c decreases linearly (0.3 < t < 0.4 ms), and finally becomes c = 200 kg/s (t > 0.4 ms).

Figure 10 shows the impact force waveform calculated by the Voigt model. The impact force waveform rises vertically at the beginning of the impact due to the damping coefficient of the dashpot. Although the thorn-shaped waveform caused by the viscosity transient appears, the thorn-shaped waveform is different from the measured waveform.

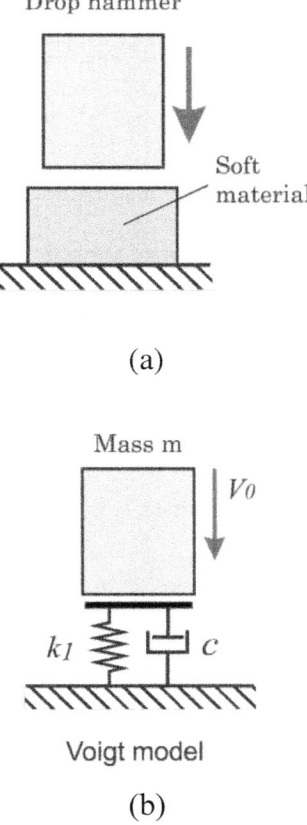

(a)

(b)

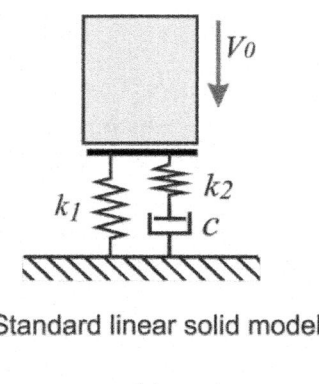

Standard linear solid model

(c)

Figure 8: Voigt model and standard linear solid model for the simulation of the impact force waveform of a soft material.

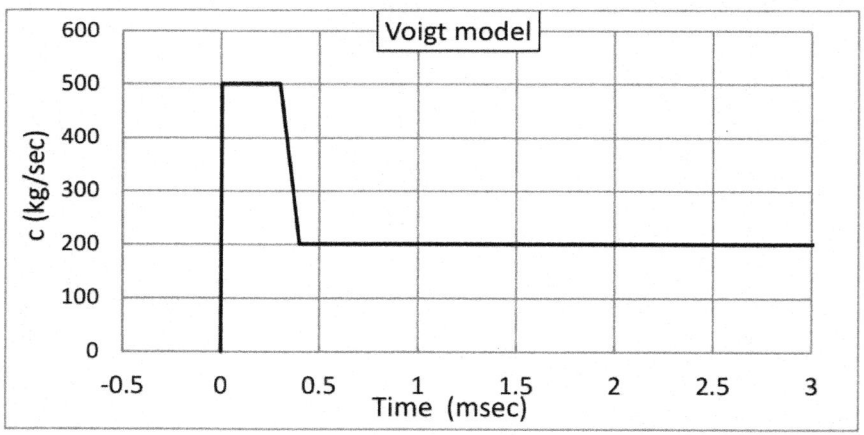

Figure 9: Damping coefficient c assumed in the Voigt model and the standard linear solid model.

Next, an analysis using the SLS model is carried out. Here, k_1 and c are the same as the values used in the above mentioned Voigt model. The value of k_2 is assumed to be 5000 kN/m. Figure 11 shows the impact force waveform calculated by the SLS model. By placing the spring k_2 in series with the damper, a small thorn-shaped waveform similar to that measured in the experiment is generated. From this figure, the SLS model is considered to be effective in the simulation of the thorn-shaped waveform.

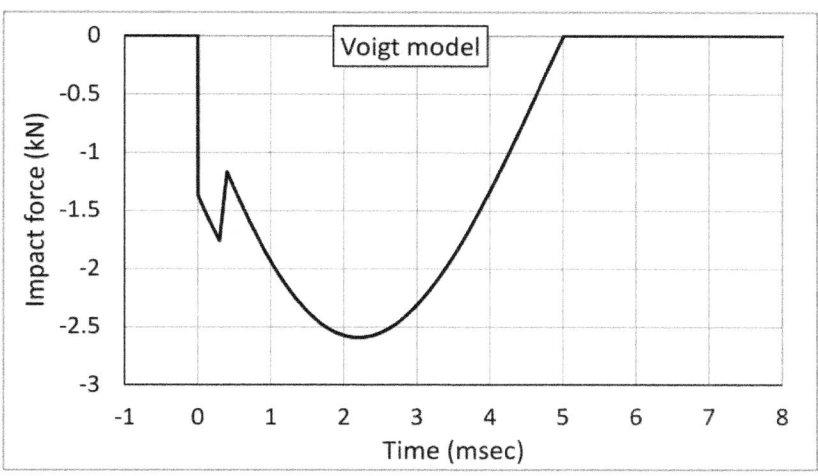

Figure 10: Impact force waveform simulated using the Voigt model.

Several types of thorn-shaped waveform are calculated using the SLS model. Figure 12 shows the impact force waveforms calculated using the SLS model for the three cases listed in Table 1. The values of k_2 and c in the table are the same for these three cases, and only k_1 is changed. The figure shows that the impact period becomes longer as k_1 is reduced.

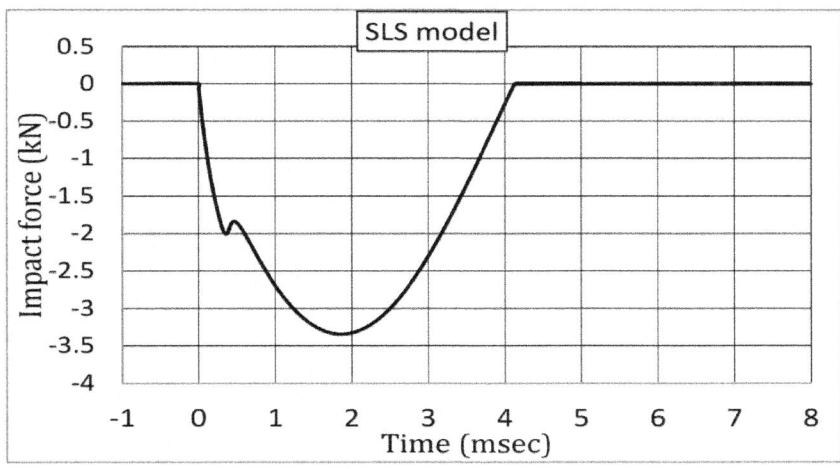

Figure 11: Impact force waveform simulated using the standard linear solid model.

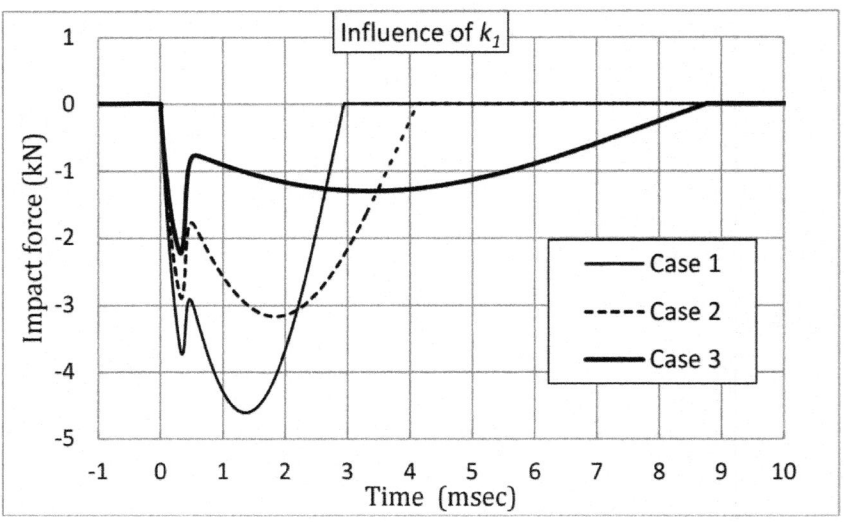

Figure 12: Impact force waveforms calculated using the standard linear solid model.

Table 1: Spring coefficients k_1, k_2 and damping coefficient c for the three cases calculated

	k_1 (kN/m)	k_2 (kN/m)	$0 < t < 0.3$ ms	$0.3 < t < 0.4$ ms	$t > 0.4$ ms
			c (kg/s)		
Case 1	2000	5000	1000	Linearly decrease	200
Case 2	1000	5000	1000	Linearly decrease	200
Case 3	200	5000	1000	Linearly decrease	200

SIMULATION OF IMPACT FORCE WAVEFORMS OF AC-TUAL SOFT MATERIALS

The SLS model is applied to the simulation of the impact force waveform of actual soft materials. Nitrile rubber is chosen as the materials for the simulation. The simulation is conducted using several values of k_1, k_2, and c in order to determine the condition in which the calculated waveform becomes close to the experimental waveform of the respective material. In other words, k_1, k_2, and c are decided such that the thorn height, the mountain height, and the total impact period of the simulated waveform match those of the experimental waveform.

Figure 13 shows the impact force waveforms of nitrile rubber simulated using the SLS model. The spring coefficients k_1 and k_2 are assumed to be constant during the impact period, and are assumed to be k_1 = 1000 kN/m and

k_2 = 3000 kN/m. The damping coefficient c of the dashpot is assumed to be a time dependent value, which changes during the impact period. Figure 14 shows the time variation of c. c = 1500 kg/s when 0 < t < 0.2 ms. During the period of 0.2 < t < 0.4 ms, c decreases linearly from 1500 kg/s to 100 kg/s. During the period of t > 0.4 ms, c = 100 kg/s. The period 0.2 < t < 0.4 ms is the period of the viscosity transient. A thorn shape is clearly observed in the rising segment of the waveform in Figure 14. The peak height of the thorn, the peak height of the mountain-shaped waveform, and the total impact period are approximately the same as those of nitrile rubber, as shown in Figure 13.

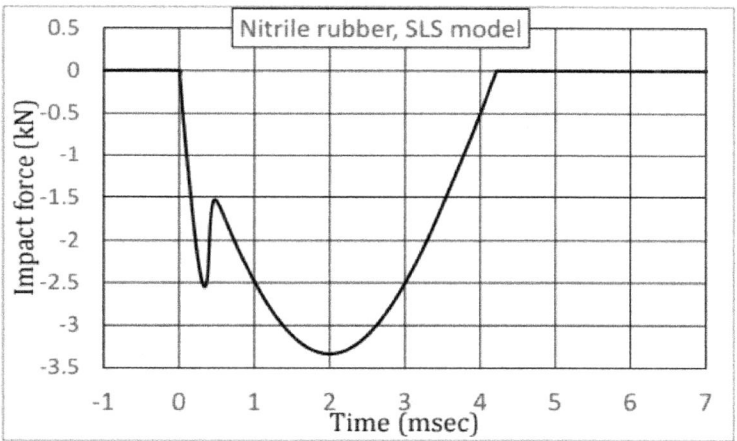

Figure 13: Impact force waveform of nitrile rubber simulated using the standard linear solid model.

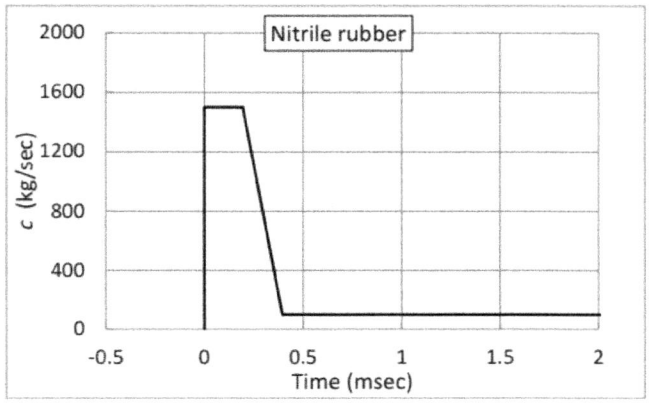

Figure 14: Damping coefficient c assumed for nitrile rubber.

CONCLUSIONS

When an impact test on soft material is carried out under the condition of a flat frontal impact, an impact force waveform consisting of a thorn-shaped waveform and a subsequent mountain-shaped waveform is obtained. We believe that this distinctive waveform is a result of the viscosity transient during the impact period. In this study, we attempt to simulate the impact force waveforms using a dynamics model. The obtained results are as follows.

1. When a soft material receives an impact force, a large viscous resistance is induced at the beginning of the impact. The viscous resistance then decreases rapidly with time. This viscosity transient phenomenon is thought to be the reason for the occurrence of the thorn-shaped waveform.

2. The standard linear solid (SLS) model can qualitatively explain the feature of the impact force waveforms of soft materials. In the SLS model, the occurrence mechanism of the thorn-shaped waveform is explained by treating the damping coefficient of dashpot as a time-dependent value.

3. A more sophisticated dynamics model is necessary in order to describe the details of the impact force waveform of the respective soft material. The densification mechanism of the material, the double-strike phenomenon, etc., should be considered in the model for the quantitative simulation.

REFERENCES

1. Fujimoto, Y., Liu, C., Tanaka, Y. and Shintaku, E. (2013) Measurement of Double-strike Phenomenon of Impulsive Force Using Fluctuating Load Detection Plate. Journal of the Japanese Society for Experimental Mechanics, 13, 112- 120.

2. Song, B., Chen, W.W., Ge, Y. and Weerasooriya, T. (2007) Radial Inertia Effects in Kolsky Bar Testing of Extra-Soft Specimens. Experimental Mechanics, 47, 659-670.http://dx.doi.org/10.1007/s11340-006-9017-5

3. Bergstorm, J.S. and Boyce, M.C. (1998) Constitutive Modeling of the Large Strain Time-dependent Behavior of Elastomers. Journal of the Mechanics and Physics of Solids, 46, 931-954. http://dx.doi.org/10.1016/S0022-5096(97)00075-6

4. Roylance, D. (2001) Engineering Viscoelasticity. 1-37.http://ocw.mit.edu/courses/materials-science-and-engineering/3-11-mechanics-of-materials-fall-1999/modules/visco.pdf

5. Fujimoto, Y., Liu, C., Uesugi, T., Tanaka, Y. and Shintaku, E. (2013) Pipe Surface Sensor for Impulsive Force Measurement. Transactions of the Japan Society of Mechanical Engineers Series C, 79, 1949-1959. http://dx.doi.org/10.1299/kikaic.79.1949

6. Hoffman, R.L. (1972) Discontinuous and Dilatant Viscosity Behavior in Concentrated Suspensions I . Observation of a Flow Instability. Transactions of the Society of Rheology, 16, 155-173. http://dx.doi.org/10.1122/1.549250

7. Bender, J. and Wagner, N. (1996) Reversible Shear Thickening in Monodisperse and Bidisperse Colloidal Dispersions. Journal of Rheology, 40, 899-915.http://dx.doi.org/10.1122/1.550767

Chapter 4

DYNAMICS OF DRAINAGE OF POWER-LAW LIQUID INTO A DEFORMABLE POROUS MATERIAL

Javed I. Siddique[1], Forrest A. Landis[2], and Muhammad R. Mohyuddin[3,4]

[1]Department of Mathematics, Pennsylvania State University, York Campus, York, USA

[2]Department of Chemistry, Pennsylvania State University, York Campus, York, USA

[3]Department of Mathematics, FAST University, Islamabad, Pakistan

[4]NCBA & E, Gujrat, Pakistan

ABSTRACT

In this study we explore the one-dimensional drainage of a power-law fluid into a deformable porous material. Initially, the fluid is imbibed into the dry undeformed material due to capillary suction which in turn deforms the porous material and forms liquid and solid interfaces. Mixture theory is employed to study the movement of the liquid and solid phases. The zero-gravity model contains the similarity solution that is solved numerically. The stress gradient within the deformable porous material is induced from a pressure gradient that produces an evolving solid fraction and hence deformation. In the absence of gravity effects, the deformation of the solid seems in the same direction of imbibition. This is because of attraction of gravity. Note that these liquid and solid dynamics depend on both the power-law indexes n and m. We performed the experiments to measure the drainage and deformations of deformable porous materials for two samples of silicon oil (polydimethylsiloxane) in a polyurethane foam. Our experiments show that the silicon with high viscosity drains slower than silicon oil with low viscosity. The theoretical and experimental results show the same qualitative trend.

INTRODUCTION

In this study we develop a model for the drainage of a power law liquid into deformable porous materials. The motivation of this work is connected to many scientific fields, such as oil recovery, inkjet printing, textile engineering, soil consolidation, reservoir engineering and biomechanics. The movement of fluid through the porous materials swells the porous material which in turn affects the flow of fluid. This alteration in fluid flow and deformation of the porous material identifies the complexity and importance of these flows. Interestingly enough, most of the fluids involved in these processes are not Newtonian in nature and hence should be incorporated in the complex dynamics of deformation. To accommodate this need, we present a coupled model for deformation of porous material and drainage of power law fluids.

An overview of literature shows that the deformation of porous materials couple with fluid flow goes back to Terzaghi [1] . This idea was later extended by Biot [2] [3] to study soil consolidation. Later on, mixture theory [4] [5] was introduced to enhance the understanding of material deformation causes by fluid imbibition. Some of the biological applications that use the mixture theories include articular cartilage [6] - [12] , arterial tissue [13] - [16] and skin [17] . Note that most of the fluids involved in the above referenced biological studies possess non- Newtonian properties and therefore, non-Newtonian properties must be incorporated with the mixture theory in future studies.

Most industrial applications (e.g. composite materials, paper and inkjet printing, and dyeing of colored fabrics) involve fluids that are non-Newtonian. In an effort to address this behavior, Sommer and Mortensen [18] studied a forced unidirectional infiltration of deformable porous materials. They considered a constant pressure driven flow in an initially dry sponge like material and an agreement between theory and experiment was reported. A similar model of an infiltration of an incompressible liquid into an initially dry porous material was developed by Preziosi et al. [19] , where they allowed the porous material to deform and relax. Another study that uses mixture theory to model the imbibition of a liquid droplet on a deformable substrate in a one dimensional setting was presented by Anderson [20] . In the absence of gravity effects, the imbibition causes swelling, swelling relaxation and shrinking of a porous material.

The history of capillary rise into porous media goes back to the classical model first presented by Washburn [21] . This model shows that the amount of liquid imbibed into a porous material is proportional to $\sqrt{t}$. To validate the Washburn model, Zhmud et al. [22] and Lago and Araujo [23] presented an experimental and theoretical model of capillary rise into porous materials. Their results are consistent with the Washburn model for initial times and deviate from this trend for long times. Following these studies, Siddique et al. [24] presented an analog of the Washburn model of capillary rise. Mixture theory has been used to take into account the deformation of porous materials.

The non-Newtonian modeling along with the mixture theory was studied by Siddique and Anderson [25] . In particular, they studied the capillary rise of a power law fluid into a deformable porous material, where they assumed imbibition occurring from an infinite bath of power law fluid. When gravity effects are present, both liquid and solid interface positions reach equilibrium heights depending upon the power law index n. This study mimics only one particular aspect of capillary rise; however the need is to extend the current power law model along with mixture theory to explore many other physical settings. In this study we explore the drainage of a finite amount of liquid into deformable porous materials in the presence and absence of gravity effects.

There are many types of non-Newtonian fluid models, e.g. Herschel-Bulkley fluid, power law and differential fluid type, etc. These fluid models have been widely used in a variety of settings to analyze various physical aspects. Most relevant ones are Christopher and Middleman [26] , Sadowski [27] and Hayes et al. [28] . These studies explained some of the important aspects of porous medium flow based on power law model but there are still important phenomena to explain and clarify. The current study is an effort to model the power law fluid with mixture theory (for details see [25]).

In this paper, we study the drainage of a power law fluid into deformable porous material. The initial height $H(t)$ of fluid is modeled via an Equation. Our experiments of drainage of silicon oil with low and high viscosity into deformable porous material encouraged us to present a mathematical model that can be used to predict the similar dynamics of drainage. We do this by following Siddique et al. [24] [25] . We hope that our preliminary efforts in this regard may lead to further investigate the additional features of flows in complex porous media.

EXPERIMENT

A sample of polyurethane foam (pore size ca. 50 - 150 mm (Figure 1)) was cut into a $1 \times 1 \times 2$ cm piece. The foam was then inserted into a 1 cm² plastic cuvette with a hole drilled in the bottom to allow for drainage. The foam fit just snuggly in the cuvette so that liquid could not flow around the foam, but little compression of the foam occurred which would alter the drainage of the liquid (Figure 2). Two samples of silicon oil (polydimethylsiloxane) were used as the drainage fluid: a low viscosity sample (10 cPs) and a high viscosity sample (1000 cPs). Since these non-Newtonian silicon oils have identical chemical structure, their molecular interaction with the foam should also be identical; however, they have different molar masses, which is manifested in their differences in viscosity. Only these differences in viscosity should affect their drainage through the foam. A single grain of solid iodine was dissolved in both silicon oils to provide contrast with the foam and aid in the determination of the movement of the oils through the foam. It is not expected that the presence of the iodine will alter the drainage of the oils. 1 mL of the dyed silicon oil was injected into the top of the cuvette and it began to drain into the foam. A digital camera operating at 30 frames per second and a resolution of 1920×1080 pixels was used to record the flow of the silicon oils through the foam. Using the images produced by the camera, the position (in pixels) of the oil above the foam, $H(t)$, the foam interface position, $h_s(t)$, and the liquid interface, $h_l(t)$, were determined as a function of time (Figure 2). The distance in pixels was converted to distance in mm using the length of the cuvette (1 cm) as a scale. The experiment was concluded when the level of the silicon above the foam reached the level of the foam.

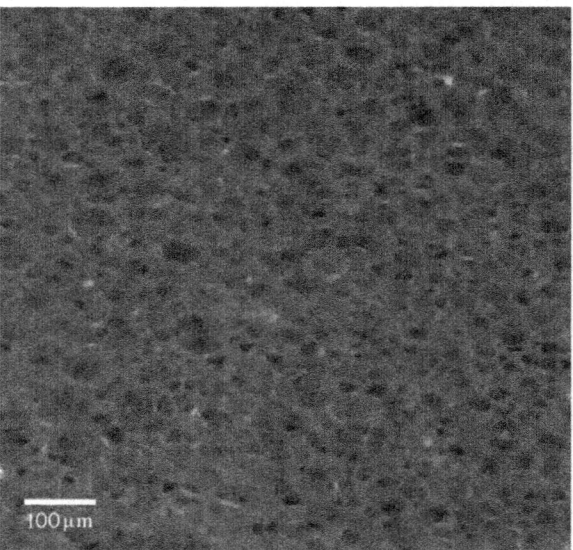

Figure 1: Optical micrograph of a cross-section of the polyurethane foam.

Experimental Results

Figure 2 shows a representative drainage experiment using the high viscosity silicon oil as the fluid. The beginning point (0 s in Figure 2) was chosen as the image where all of the 1 mL volume of silicon oil had been added to the cuvette. As the experiment continued, the height of the liquid above the foam, $H(t)$, and the liquid interface position, $h_i(t)$, both dropped as the oil absorbed into the foam (10 s and 30 s in Figure 2). This experiment was terminated when all of the silicon oil had penetrated into the foam (when $H(t) = h_s(t)$ at ca. 100 s). The positions of each of the three interfaces are plotted as a function of time in Figure 3 for both the low and high viscosity silicon oils. It took approximately 4 s for the low viscosity oil to completely drain into the foam while the higher viscosity oil took significantly longer at 100 s.

It should be noted that no measurable expansion of the foam was observed (i.e., $h_s(t)$ was constant) for these foams when imbibed with either silicon oil. This is in contrast to drainage experiments where low molar mass alcohols such as ethanol were used. While ethanol has similar rapid drainage characteristics to the low viscosity silicon oil, it differed in that it also caused a significant expansion of the foam. Ethanol is a polar liquid that can interact strongly with the polyurethane foam causing expansion. In contrast, the silicon oil is essentially nonpolar and will not interact strongly with the foam resulting in little expansion. Clearly the expansion of the foam is very dependent on the nature of the intermolecular interactions between the foam and the liquid and will be examined in future research efforts.

Figure 2: Representative photographs showing the drainage of the high viscosity silicon oil into the polyurethane foam at different time intervals. The liquid height above the foam, the foam height, and the liquid interface are labeled ($H(t)$, $h_s(t)$, and $h_i(t)$, respectively).

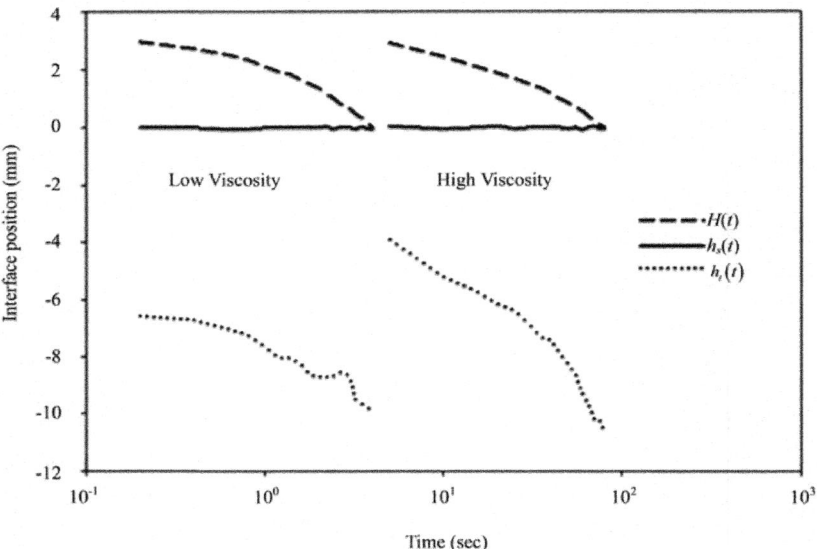

Figure 3: This figure shows the dynamics behavior where the curves are shown in dimensional form. The dashed curves shows the initial height as a function of time $H(t)$, the solid line shows the deformation of foam as function of time $h_s(t)$ and drainage of liquid as a function of time $h_l(t)$.

Figure 3 shows the dynamics behavior of the low and high viscosity silicon oils into the polyurethane foam. Here the curves are shown in dimensional form. The dashed curves shows the initial height as a function of time $H(t)$, the solid line shows the deformation of foam as function of time $h_s(t)$ and drainage of liquid as a function of time $h_l(t)$.

MATHEMATICAL MODELING

The basic geometrical description for our problem is shown in Figure 4. We consider a finite amount of a non-Newtonian liquid with an initial height defined as H_0 in contact with a deformable porous material at time $t = 0$. In the figure this initial contact position between the liquid and the deformable porous material is specified by $z = 0$. We assume the pressure at the positions $z = H(t)$ is atmospheric pressure. Later, the height of the liquid will be represented by the Equation for $H(t)$. After time $t > 0$, the non-Newtonian liquid starts to imbibe into the deformable porous material which in turn deforms the porous material and forms two interfaces: the upper interface $z = h_s(t)$, and lower interface $h_l(t)$

, as can be seen from Figure 4. In order to define the dimensionless system of equations, we use the following set of scaling parameters

$$z' = \frac{z}{L}, \quad t' = \frac{t}{T}, \quad u'_s = \frac{T}{L}u_s, \quad u'_1 = \frac{T}{L}u_1, \quad p' = \frac{p}{\Sigma_0},$$

$$h'_s = \frac{h_s}{L}, \quad h'_1 = \frac{h_1}{L}, \quad \sigma'(\phi) = \frac{\sigma(\phi)}{\Sigma_0}, \quad K'(\phi) = \frac{K(\phi)}{K_0}.$$

The above dimensionless quantities suggest the following time and length scales after balancing the terms in both momentum equations $T^n = \frac{L^{n+1}\mu^*}{md^{n+1}}$ and $L = \frac{m\rho_1}{g}$. Note that this choice of time scale depends on power-law index n. We will present the details on how to make this choice of time scale independent of power law index n for numerical simulation in the result section. In the above defined set of scaling parameters we use Σ_0 and K_0 are representative scales for solid stress and the permeability respectively. We follow Siddique and Anderson [25] where they defined the permeability $K(\phi)$ in terms of the permeability of bed particles of diameter d and solid volume fraction ϕ (see Hayes et al. [28] and references therein for details) and μ_{eff} for the power law fluid is given by

$$K(\phi) = W(\phi)d^2, \quad W(\phi) = \frac{(1-\phi)^3}{150\phi^2} \tag{1}$$

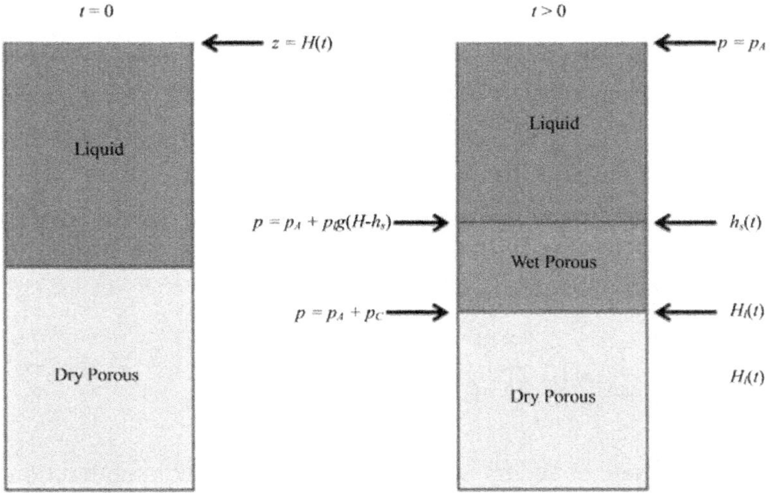

Figure 4: Problem configuration at $t = 0$ and $t > 0$.

$$\mu_{eff} = \mu^* Y(\phi,n) d^{1-n}$$

(2)

where.

Note, setting $n=1$ in $Y(\phi,n)$ relation and Equation (2) yields $\mu_{eff} = \mu^*$, that represents the Newtonian viscosity. The resulting set of Equations describing the one-dimensional drainage of non-Newtonian liquid (e.g. power law) into deformable porous material after dropping the primes is

$$\phi_t + (\phi u_s)_z = 0$$

(3)

$$\phi_t - [(1-\phi)u_1]_z = 0$$

(4)

$$(u_1 - u_s)^{n-1}(u_1 - u_s) = -\frac{K(\phi)}{(1-\phi)}(p_z + 1)$$

(5)

$$p_z = \sigma_z - (\rho\phi + 1),$$

(6)

where $\rho = \rho_s^T / \rho_l^T - 1$ and stress is a function of solid volume fraction $\sigma = \sigma(\phi)$. Following Anderson 2005, we assume $\sigma(\phi) = \phi_r - \phi$. This choice is suitable for one dimensional deformation and needs to be generalized to take into account effects such as shear deformation in a higher dimension. It is worthwhile mentioning that subscript t and z represent the derivative with respect to corresponding variables.

Equations (3) and (4) represent mass balances and Equations (5) and (6) are momentum balances for the solid and liquid phases (see [25]). The set of parameters in these Equations can be divided into two categories, first set of parameters inside the wet material regions such as the solid volume fraction ϕ, the liquid u_1 and solid u_s component of velocities, the liquid pressure p, the solid stress $\sigma(\phi)$, and the permeability of porous material and second set represents liquid h_l and solid h_s boundary positions.

We follow the same procedure as was followed in Anderson (2005), Siddique et al. [24] , and Siddique and Anderson [25] to obtain the partial differential equation for the solid volume fraction ϕ in the wet sponge region $h_l \le z \le h_s$

$$\phi_t + c(t)\phi_z = \partial_z \left[-\frac{W(\phi)\phi^n(1-\phi)^{n-1}}{Y(\phi,n)}(\sigma'(\phi)\phi_z - \rho\phi) \right]^{1/n}$$

(7)

The appropriate boundary conditions for the solid volume fraction are

$$\phi = \phi_r, \quad \text{at} \quad z = h_s$$

(8)

$$\phi = \phi_1^* + H - h_s + \left(h_s - h_1\right)\int_0^1 (\rho\phi + 1)\,dz \quad \text{at } z = h_1.$$

(9)

In the derivation of boundary condition (8), we integrate Equation (6) after substituting $\sigma(\phi)$ and applying the pressure boundary conditions given below in (12) and (15). The liquid height Equation is derived using a conservation of liquid argument (see Siddique et al. [24])

$$H(t) = H_0 + h_s - \left(h_s - h_1\right)\int_{h_1}^{h_s} (1 - \phi)\,dz$$

(10)

where H_0 represents the initial height of the non-Newtonian liquid before it starts to drain into deformable porous material.

The boundary conditions at the liquid-wet material interface $z = h_s(t)$ are

$$u_s\left(h_s^-, t\right) = \frac{\partial h_s}{\partial t},$$

(11)

$$p\left(h_s^-, t\right) = \frac{p_A}{\Sigma_0} - \left[h_s(t) - H(t)\right]$$

(12)

$$\sigma\left(h_s^-, t\right) = 0$$

(13)

where p_A/Σ_0 is the dimensionless constant atmospheric pressure. Here, Equations (11)-(13) represent the kinematic, hydrostatic, and zero stress conditions respectively. While modeling hydrostatic boundary condition (12) we have neglected the inertial effects.

The kinematic and pressure boundary conditions at wet-material-dry material interface $z = h_1(t)$ are

$$u_1\left(h_1^+, t\right) = \frac{\partial h_1}{\partial t},$$

(14)

$$p\left(h_1^+, t\right) = p_A/\Sigma_0 + p_c/\Sigma_0$$

(15)

where p_c/Σ_0 represents the dimensionless constant capillary pressure.

Boundary conditions (11) and (14) transform to ordinary differential Equations for the solid and liquid interfaces

$$\frac{dh_s}{dt} = c(t) - \left[\frac{(1-\phi)^{n-1} W(\phi)}{Y(\phi, n)}\left(\frac{\partial \phi}{\partial z} + \rho\phi\right)\bigg|_{z=0}\right]^{\frac{1}{n}}$$

(16)

$$\frac{dh_1}{dt} = c(t) + \left[\frac{W(\phi)\phi^n}{Y(\phi,n)(1-\phi)} \left(\frac{\partial\phi}{\partial z} + \rho\phi \right) \Big|_{z=1} \right]^{\frac{1}{n}}$$

(17)

where $c(t)$

$$c(t) = \frac{1-\phi_0}{\phi_0} \left[\frac{W(\phi)\phi^n}{Y(\phi,n)(1-\phi)} \left(\frac{\partial\phi}{\partial z} + \rho\phi \right) \Big|_{z=1} \right]^{\frac{1}{n}}$$

(18)

If we take $n=1$ in Equation (7) we recover Equation (9) of Siddique et al. [24] in dimensionless form and if we take $n=1$ and $g=0$ in (7) we recover Equation (20) of Anderson [20] in dimensionless form and Equation (44) of Prezoisi et al. [5] . The steady state solution of the above system is the same as for the Newtonian case (see Siddique et al. [24]). Below we will summarize the solution procedure in the absence of gravity and in the presence of gravity effects.

GRAVITY INDEPENDENT SOLUTION

In the absence of gravity effects, Equation (7) admits the solution in terms of a similarity variable

$\eta = \dfrac{z}{(n+1/n)t^{\frac{n}{n+1}}}$ that yields the following ordinary differential Equation

$$\left(-\frac{n}{n+1} \right)^{1/n} \eta \frac{d\phi}{d\eta} + \frac{1-\phi_0}{\phi_0} \left[\frac{W(\phi)\phi^n}{Y(\phi,n)(1-\phi)} \frac{d\phi}{d\eta} \Big|_{\lambda^+} \right]^{\frac{1}{n}} \frac{d\phi}{d\eta} = \frac{d}{d\eta} \left[\frac{W(\phi)\phi^n(1-\phi)^{n-1}}{Y(\phi,n)} \frac{d\phi}{d\eta} \right]$$

(19)

subject to the boundary conditions $\phi(\lambda_s) = \phi_r,$ and $\phi(\lambda_1) = \phi_1^*$, where $\phi_1^* = \phi_r - p_c/\Sigma_0$. The boundary condition, $\phi(\lambda_1) = \phi_1^*$ is obtained by substituting $\sigma = \sigma(\phi)$ in Equation (6), integrating, and using boundary con- ditions (12) and (15) in the absence of gravity. Note that λ_s and λ_1 represents the interface positions in terms of the similarity variables given below. Similarly, introducing $h_s(t) = (n+1/n)\lambda_s t^{n/n+1}$ and $h_1 = (n+1/n)\lambda_1 t^{n/n+1}$ yields this following relations for λ_s and λ_1

$$\lambda_s = \left(\frac{n}{n+1} \right)^{1/n} \left[\frac{1-\phi_0}{\phi_0} \left(\frac{\phi^n W(\phi)}{(1-\phi)Y(\phi,n)} \frac{d\phi}{d\eta} \Big|_{\lambda_1^+} \right)^{1/n} - \left(\frac{(1-\phi)^n W(\phi)}{Y(\phi,n)} \frac{d\phi}{d\eta} \Big|_{\lambda_s^-} \right)^{1/n} \right]$$

(20)

$$\lambda_1 = \left(\frac{n}{n+1}\right)^{1/n} \left[\frac{1}{\phi_0}\left\{\frac{\phi^n W(\phi)}{(1-\phi)Y(\phi,n)}\frac{d\phi}{d\eta}\bigg|_{\lambda_1^+}\right\}^{1/n}\right] \tag{21}$$

The non-linear ODE (19) along with the non-linear Equations (20) and (21) is solved numerically. The ODE and non-linear Equations are discretized using finite difference and midpoint rule yielding a system of non-linear Equations. This coupled system of non-linear Equations is solved numerically. We will present the solution of zero gravity case in the results section. It is worthwhile highlighting that this numerically computed zero gravity solution will be used as an initial condition for non-zero gravity case which we will discuss in the section below.

NON-ZERO GRAVITY SOLUTION

In the presence of gravity effects, we will first transform the moving domain problem given by (7) along with (16) and (17) to fixed domain using the following transformation

$$z = \frac{z - h_1(t)}{h_s(t) - h_1(t)}. \tag{22}$$

This helps transforming the moving domain problem $h_1 \le z \le h_s$ to a fixed domain problem $0 \le z \le 1$. We use $h_s(t=0)=0$ and $h_1(t=0)=0$ initial conditions for solid and liquid interface positions respectively. We use method of lines along with boundary conditions given in (8) and (9). Again, we use the zero gravity solution as initial conditions for the non-zero gravity case. We approximate the spatial derivatives using second order ac- curate finite difference and mid-point discretizations. This reduces the system of partial differential Equations (PDEs) to system of ordinary differential Equations (ODEs) in time which is solved using Matlab's solver ode23s. We start our numerical integration for $0 < t < t_{start}$, where t_{start} is a numerical small value.

RESULTS

Our results are based on a specific set of test fluids for which power-law index n and consistency index μ^* values are available (see Missirlis et al. [29]). The calculation of the μ^* values is based on the assumption that the capillary pressure $|p_c| = (\gamma\cos\theta)/d$, where γ is the surface tension assumed to be same and non-New- tonian case and θ is the wetting angle, that we also assumed to be $\theta = 0$. We introduce the Newtonian time scale $T = \mu_N/|p_c|$ in such a way that computed simulation results are independent of the power law index n. Note that the time scales are related through capillary pressure p_c that depends on particle diameter d. When capillary pressure increases the drainage of fluid increases. Although we do not explore the dependence on μ^* here, increasing μ^* means increasing friction force. In other words, μ^* is directly proportional to the drainage process.

In the beginning of the process, a finite amount of liquid is supplied, whose height is shown by $H(t)$ in Figure 5. It is important to note that when gravity effects are not present, the solution is independent of p, and both interface positions $h_s(t)$ and $h_1(t)$ admits similarity solution until the entire fluid is drained into the deformable porous material (i.e., $h_s(t) = H(t)$). Figure 5summarizes the drainage comparison between the Newtonian and different power law fluids. Each of these curves corresponds to set of parameters $(\phi_0, \phi_r, \phi_1^*)$ used in sponge experiment in Siddique et al. [24] along with power law consistency index μ^* and the power law index n values from Missirlis et al. [29] . However it is important to note that equilibrium heights $h_s(t)$ and $h_1(t)$ attained depend on the values of ϕ_0, ϕ_r, ϕ_1^* and p. The bottom right of Figure 5 shows power law index n as a function of t. As n decrease the drainage time increases. We denote this drainage time as t^*. It is interesting to note that both the Newtonian and non-Newtonian fluids follow the similarity behavior but as n decreases, t^* increases which means decreasing the n values slows down the dynamics. Figure 6 shows the drainage of liquid into a deformable porous material in the presence of gravity effects. The time scales used in Figure 6 allow us to present a direct comparison for different values of n. Similar to the Newtonian case, a finite amount of liquid is supplied, shown by $H(t)$ in Figure 6. For this case, both curves $h_s(t)$ and $h_1(t)$ follow similarity solution for very short time and then depart from this behavior until the fluid is entirely drained into porous material. The drainage time for the nonzero gravity case is faster than the zero gravity case.

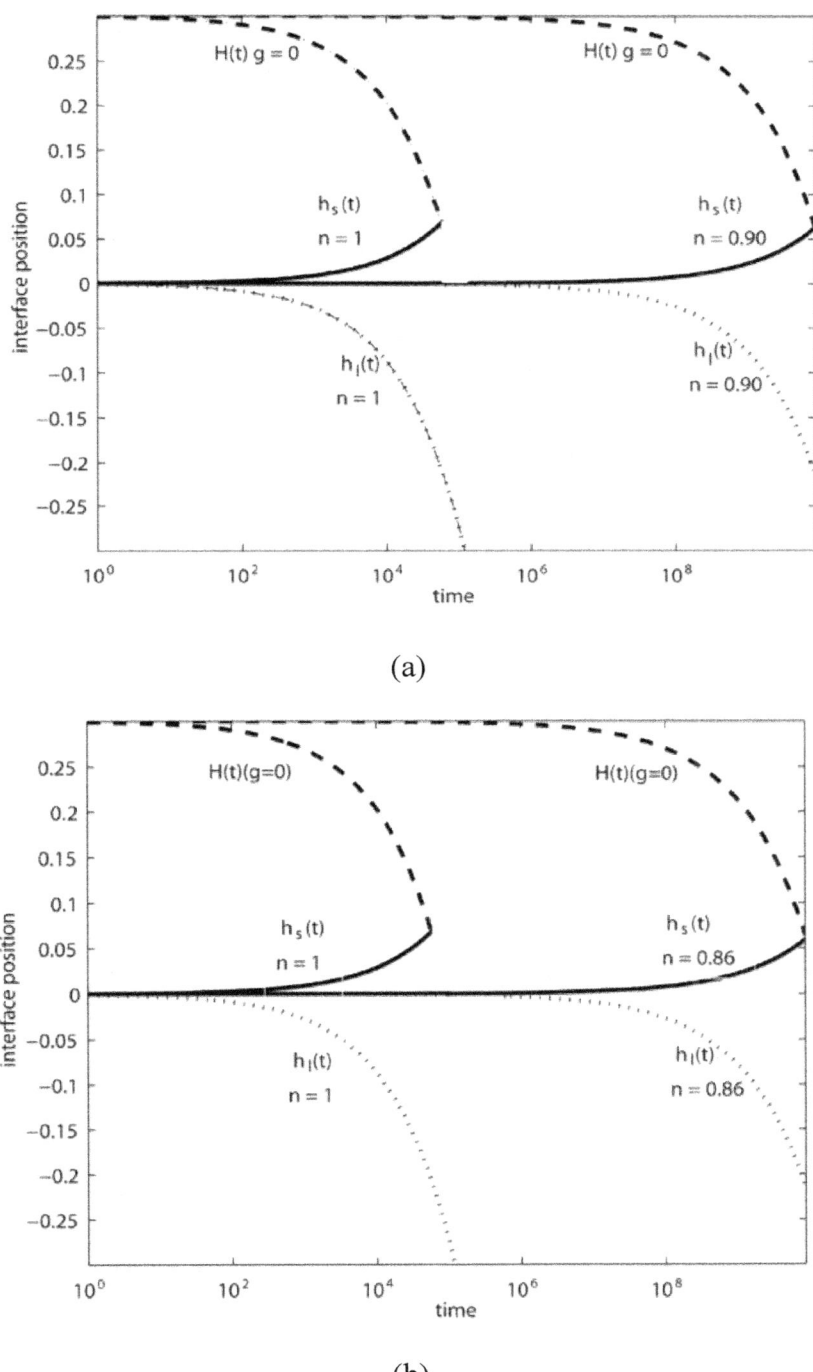

(a)

(b)

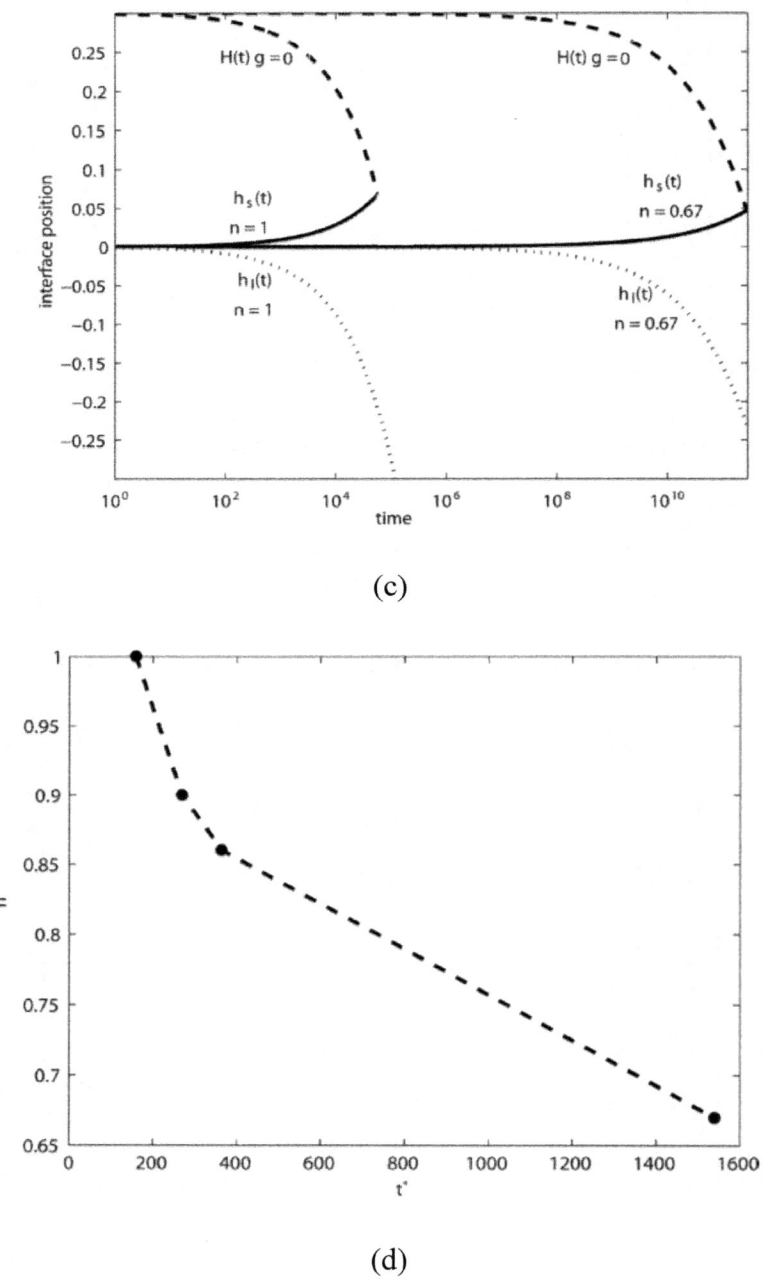

(c)

(d)

Figure 5. The plots show the drainage of Newtonian and non-Newtonian liquids. Top left shows the comparison between $n = 1$ and $n = 0.90$, top right $n = 1$ and $n = 0.86$ and bottom left $n=1$ and $n=0.67$. Bottom right plot shows that drainage time t^* for different for different fluids. The other parameter used in these figures are $\rho_1 = 1000 \text{ kg/m}^3$, $g = 9.8 \text{ m/sec}^2$, $d = 0.001$, $\phi_0 = 0.33$, and surface tension $63.57 \times 10^{-3} \text{ N/m}$.

In order to connect our theoretical predictions with experiments, we performed drainage of silicon oil of two different viscosities with identical chemical structure into deformable sponge like materials. Our experimental predictions show a similar trend for drainage and deformation of deformable porous materials. The drainage of water into deformable porous material follow approximately $t^{1/2}$ which is expected for the case of Newtonian fluids. Our experimental curves shows a modified power law of $t^{0.48}$ and $t^{0.70}$ for two different set of silicon oils which suggests $n=0.92$ and $n=2.3$ respectively. Although we are unable to present a direct qualitative comparison between theory and experiments due to limited access to experimental facilities but our finding shows that any fluid having different properties than water require different power law behavior. We hope our findings with these preliminary comparison will encourage further experimental investigation leading to capture the realistic drainage phenomena.

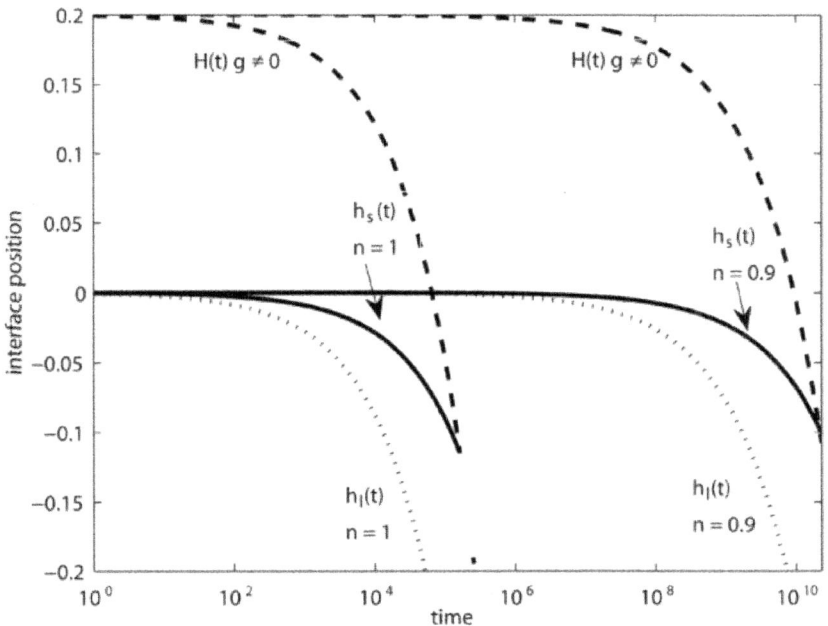

(a)

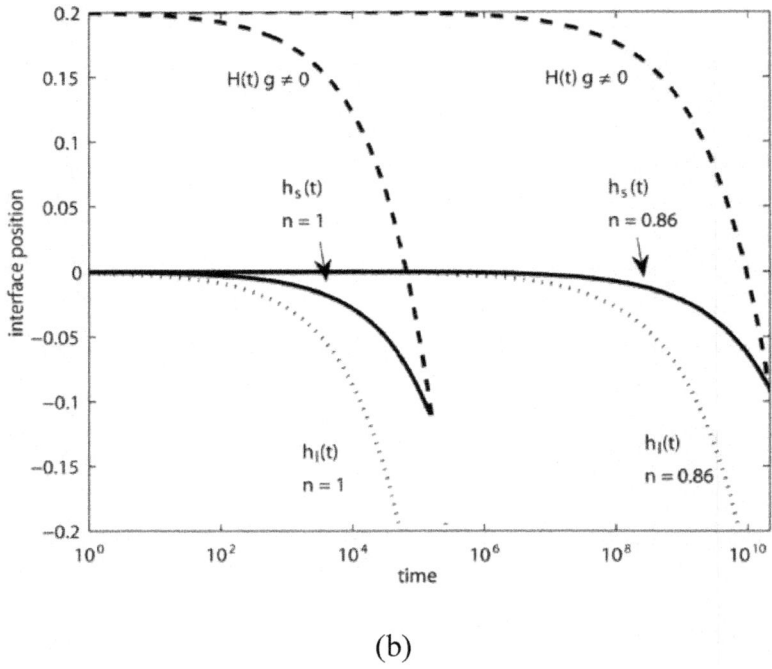

(b)

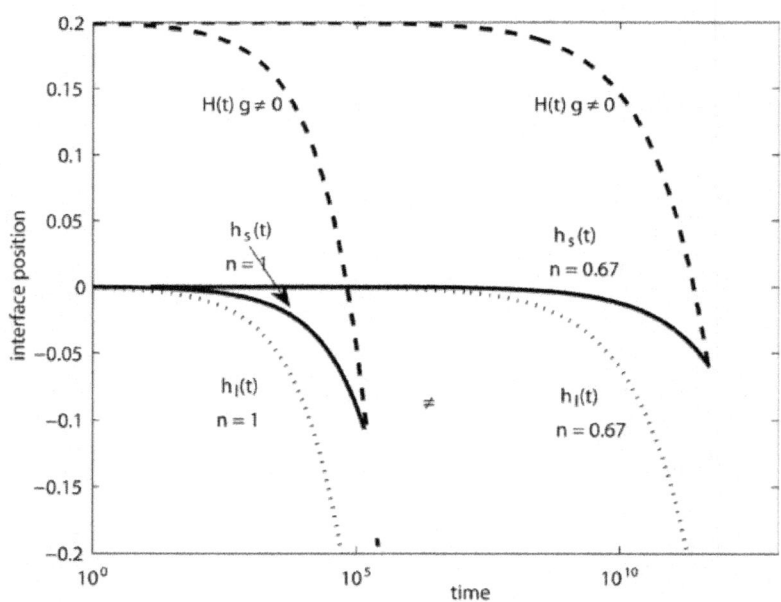

(c)

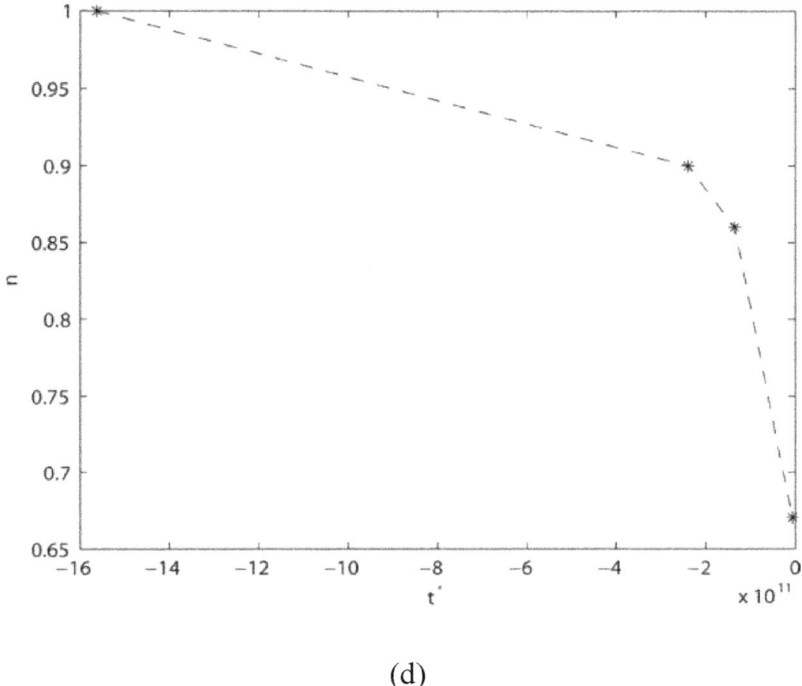

(d)

Figure 6: This plot shows the comparison of drainage of Newtonian fluid with non-Newtonian fluid for different power-law index. The curves as labeled $H(t)$, $h_s(t)$ and $h_l(t)$ for hight of liquid, height of solid, and height of liquid. Drainage time is noted as the time it takes for all of the above liquid to drain into the deformable porous material.

DISCUSSION

In this present work we have presented a basic model for predicting the drainage of a power law liquid into a deformable porous material. More specifically, we transform the capillary rise theory developed for power law liquids into deformable porous materials [24] to the drainage setting by specifying liquid height [25] via an equation based on conservation of liquid argument. In addition to theoretical predictions, this work includes a set of experiments showing the similar behavior as shown in our numerically simulations. In this study we use the mixture theory [15] [16] [18] - [20] [24] [25] to model the problem of drainage of power law liquid into deformable porous material.

In the absence of gravity effects, the deformation of the porous material and the penetration of liquid follows a different power law behavior $t^{\frac{n}{n+1}}$ compared to the Newtonian case. Our findings show that the dynamics for both h_s and h_1 follow different dynamics compared to the Newtonian case. We were able to compute the drainage time for different fluids. These findings are consistent with our experimental findings. When gravity effects are present in our model, the drainage of power law liquid depends on ρ. Note that both curves, $h_s(t)$ and $h_1(t)$, follow similarity solution initially and then depart from this trend depending upon their power law consistency index μ^* and power law index n. In short, the dynamics for fluid with different properties follow different dynamics.

The present study is preliminary, but suggests some interesting possibilities in the theory of drainage of power law fluid. An interesting area to explore in future investigations would be the use of a more generalized function for solid stress σ and a complicated permeability function $K(\phi)$ and for exploring the deformation in higher dimension. Alternatively, keeping a simplified functions for σ and $K(\phi)$ and extending the ability to directly compare the theoretical and experimental predictions could lead to a better understanding of the basic mechanisms of drainage

ACKNOWLEDGEMENTS

The author J. I. Siddique greatly appreciate the support of Simons Foundation Grant No. 281839.

REFERENCES

1. Terzaghi, K. (1925) Erdbaumechanik auf Bodenphysikalischen Grundlagen. Deuticke, Wien.

2. Biot, M.A. (1955) Theory of Elasticity and Consolidation for a Porous Anisotropic Solid. Journal of Applied Physics, 26, 182-185. http://dx.doi.org/10.1063/1.1721956

3. Biot, M.A. (1962) Mechanics of Deformation and Acoustic Propogation in Porous Media. Journal of Applied Physics, 33, 1482-1498.

4. Atkin, R.J. and Crain, R.E. (1976) Continuum Theories of Mixture: Basic Theory and Historical Development. Quarterly Journal of Mechanics and Applied Mathematics, 29, 209-244. http://dx.doi.org/10.1093/qjmam/29.2.209

5. Bowen, R.M. (1980) Incompressible Porous Media Models by Use of the Theory of Mixtures. International Journal of Engineering Science, 18, 1129-1148.http://dx.doi.org/10.1016/0020-7225(80)90114-7

6. Lai, W.M. and Mow, V.C. (1980) Drag Induced Compression of Articular Cartilage during a Permeation Experiment. Biorheology, 17, 111-123.

7. Holmes, M.H. (1983) A Nonlinear Diffusion Equation Arising in the Study of Soft Tissue. Quarterly of Applied Mathematics, 41, 209.

8. Holmes, M.H. (1984) Comparison Theorems and Similarity Solution Approximations for a Nonlinear Diffusion Equation Arising in the Study of Soft Tissue. SIAM Journal on Applied Mathematics, 44, 545-556. http://dx.doi.org/10.1137/0144037

9. Holmes, M.H. (1985) A Theoretical Analysis for Determining the Nonlinear Hydraulic Permeability of a Soft Tissue from a Permeation Experiment. Bulletin of Mathematical Biology, 47, 669-683. http://dx.doi.org/10.1007/BF02460132

10. Holmes, M.H. (1986) Finite Deformation of Soft Tissue: Analysis of a Mixture Model in Uni-Axial Compression. Journal of Biomechanical Engineering, 108, 372-381.http://dx.doi.org/10.1115/1.3138633

11. Holmes, M.H. and Mow, V.C. (1990) The Nonlinear Characteristic of Soft Gels and Hydrated Connective Tissue in Ultrafiltration. Journal of Biomechanics, 23, 1145-1156.http://dx.doi.org/10.1016/0021-9290(90)90007-P

12. Hou, J.S., Holmes, M.H., Lai, W.M. and Mow, V.C. (1989) Boundary Conditions at the Cartilage-Synovial Fluid Interface for Joint Lubrication and Theoretical Verifications. Journal of Biomechanical Engineering, 111, 78-87. http://dx.doi.org/10.1115/1.3168343

13. Kenyon, D.E. (1976) The Theory of an Incompressible Solid-Fluid Mixture. Archive for Rational Mechanics and Analysis, 62, 131-147.

14. Klanchar, M. and Tarbell, J.M. (1987) Modelling Water Flow through Arterial Tissue. Bulletin of Mathematical Biology, 49, 651-669. http://dx.doi.org/10.1007/BF02481766

15. Barry, S.I. and Aldis, G.K. (1992) Flow Induced Deformation from Pressurized Cavities in Absorbing Porous Tissues. Bulletin of Mathematical Biology, 54, 977-997.http://dx.doi.org/10.1007/BF02460662

16. Barry, S.I., Parker, K.H. and Aldis, G.K. (1991) Fluid Flow over a Thin Deformable Porous Layer. Journal of Applied Mathematics and Pysics (ZAMP), 42, 633-648.

17. Oomens, C.W.J., Van Campen, D.H. and Grootenboer, H.J. (1987) A Mixture Approach to the Mechanics of Skin. Journal of Biomechanics, 20, 877-885.http://dx.doi.org/10.1016/0021-9290(87)90147-3

18. Sommer, J.L. and Mortensen, A. (1996) Forced Unidirectional Infiltration of Deformable Porous Media. Journal of Fluid Mechanics, 311, 193-217. http://dx.doi.org/10.1017/S002211209600256X

19. Preziosi, L., Joseph, D.D. and Beavers, G.S. (1996) Infiltration of Initially Dry, Deforamable Porous Media. International Journal of Multiphase Flow, 22, 1205-1222.http://dx.doi.org/10.1016/0301-9322(96)00035-3

20. Anderson, D.M. (2005) Imbibition of a Liquid Droplet on a Deformable Porous Substrate. Physics of Fluids, 17, Article ID: 087140. http://dx.doi.org/10.1063/1.2000247

21. Washburn, E.W. (1921) The Dynamics of Capillary Flow. Physical Review, 17, 273-283.http://dx.doi.org/10.1103/PhysRev.17.273

22. Zhmud, B.V., Tiberg, F. and Hallstensson, K. (2000) Dynamic of Capillary Rise. Journal of Colloid and Interface Science, 228, 263-269. http://dx.doi.org/10.1006/jcis.2000.6951

23. Lago, M. and Araujo, M. (2001) Capillary Rise in Porous Media. Journal of Colloid and Interface Science, 234, 35-43. http://dx.doi.org/10.1006/jcis.2000.7241

24. Siddique, J.I., Anderson, D.M. and Bondarev, A. (2009) Capillary Rise of Liquid into Deformable Porous Material. Physics of Fluids, 21, Article ID: 013106.http://dx.doi.org/10.1063/1.3068194

25. Siddique, J.I. and Anderson, D.M. (2011) Capillary Rise of Non-Newtonian Liquid into Deformable Porous Material. Journal of Porous Media, 14, 1087-1102.http://dx.doi.org/10.1615/JPorMedia.v14.i12.40

26. Christopher, R.H. and Middlemen, S. (1965) Power-Law Flow through a Packed Tube. Industrial & Engineering Chemistry Fundamentals, 4, 422-426.http://dx.doi.org/10.1021/i160016a011

27. Sadowski, T.J. (1963) Non-Newtonian Flow through Porous Media. Ph.D. Thesis, University of Wisconsin, Madison.

28. Hayes, R.E., Afacan, A., Boulanger, B. and Shenoy, A.V. (1996) Modeling the Flow of Power Law Fluids in a Packed Bed Using a Volume-Averaged Equations of Motion. Transport in Porous Media, 41, 175-196.

29. Missirlis, K.A., Assimacopoulos, D., Mitsoulis, E. and Chhabra, R.P. (2001) Wall Effects for Motion of Spheres in Power-Law Fluids. Journal of Non-Newtonian Fluid Mechanics, 96, 459-471.

Chapter 5

THE LIGHT AS COMPOSED OF LONGITUDINAL-EXTENDED ELASTIC PARTICLES OBEYING TO THE LAWS OF NEWTONIAN MECHANICS

Alfredo Bacchieri
University of Bologna, Bologna, Italy

ABSTRACT

It is shown that the speed of longitudinal-extended elastic particles, emitted during an emission time T by a source S at speed u (escape speed toward the infinity due to all the masses in space), is invariant for any Observer, under the Newtonian mechanics laws. It is also shown that a cosmological reason implies the light as composed of such particles moving at speed u (function of the total gravitational potential). Compliance of c with Newtonian mechanics is shown for Doppler effect, Harvard tower experiment, gravitational red shift and time dilation, highlighting, for each of these subjects, the differences versus the relativity.

INTRODUCTION

Here we present a solution, in accordance to the Newtonian mechanics, to the apparent constancy of c, based on following assumptions:

1. Gravity fields fixed to their related masses (intending that each field is moving together with its generating mass).

2. Finite mass of the universe, implying a finite value of U (total gravitational potential) and therefore of u (escape speed from the universe due to all the masses in space).

3. Light composed of longitudinal-extended elastic particles (as defined on §4) moving at speed c = u. This equality is supported by a cosmological reason, see §2.

On above bases (including, needless to say, Newton's absolute time and space) we find:

- The relation between u (total escape speed) and U (total gravitational potential), giving to the speed of light the cosmological reason of its value.

- On Earth, the variation of u, (and therefore of c as per assumption III), due to the variation of U (mainly caused by the variable distance Earth-Sun) is, during one year, Δu (=Δc) $\leq \pm 0.05$ m$\times$s^{-1}, hence within the accuracy of the measured value of c.

- The invariance of the measure of c for any Reference frame under the Newtonian mechanics laws.

- The longitudinal, generic and transverse Doppler effect for longitudinal-extended elastic particles, as defined, and their physical characterization.

- As for the Harvard tower experiment [1] -[3] , regarding the variation of frequency (or wavelength) between a source (of gamma rays) and an absorber at different height, our relations give a shift equal to the observed and also predicted by the Relativity. Anyhow, with the source on the base (of the tower) the light arriving to the top has, as for the GR, a lower frequency, whereas on our bases, is the length of our particles which decreases (together with c); on the contrary, with source on the top, GR predicts an increase of the frequency of the light arrived to the base, whereas we show that, during the same path (top-base), is the length of these particles which increases (together with c), giving a red shift. Moreover, as for the value of the compensating speed source-absorber, (necessary to restore their resonance), we point out that the experiment did not give any clear indication about the effective direction of this speed. Indeed, scope of that experiment was to "establish the validity of the predicted gravitational red shift" [2] , hence the only value of this speed was taken in consideration; here, on §6, we show that, on our bases, the effective direction of this speed is contrary in both cases (source on top or base), to the one predicted (but not verified) by the Relativity.

- As for the gravitational time dilation, on §6, it is shown that taking a source (of light) in altitude, it yields a negative variation of c as well as a negative variation of the frequency n inducing atomic clocks to run faster; moreover, through our Equation (29) regarding a source circling (around the Earth), we obtain, see (46), the exact variation of the ticking time of GPS system.

- As for high red shifts related to far sources, we show that, disregarding the relative motion Earth-source, they depend on the increase of c (as well as the increase of the length of the said "longitudinal-extended elastic particles") during the path of light toward higher (in absolute value) potential; on §7, Table 1, we give the values of c (on these far sources) related to the observed red shifts.

- Our Equation (17), (regarding our Doppler effect for the light), applied to the Compton effect (indubitable Doppler effect), gives, see Appendix A, the Compton equation, which cannot be obtained through the relativistic Doppler effect equations.

TOTAL ESCAPE SPEED (FROM A POINT TOWARD THE INFINITY) DUE TO ALL THE MASSES IN SPACE

As known, considering in space one only mass M (regarded as a point-like), the gravitational potential U acting on a particle having mass m = M, assuming $U_\Psi = 0$, with s the distance M-m, is U = −MG/s; this relation, according to our first assumption (I), is always valid in spite of any reciprocal motion between M and m. The related Conservation of Energy (CoE), E = U + K, (where K $= \frac{1}{2}u^2$ represents the unitary, i.e. for unit of masskinetic energy of our particle arriving from the infinity, where $u_\Psi = 0$), for E = 0 gives U = −K, leading to

$$u = \sqrt{2K} = \sqrt{-2U} = \sqrt{2MG/s} \tag{1}$$

which is a scalar, (called escape speed), representing (in the considered point) the value of the velocity u, any massive particle, under a potential U, needs to reach the infinity, so u (escape velocity)must be referred to M.

Considering now two masses M_1 and M_2, having, at a given time, distances s_1 and s_2 from a considered point (we may call it Emission point E_p), the potential $U_{1,2}$ in E_p becomes

$$U_{1,2} \equiv U_1 + U_2 = -\left(M_1 G/s_1\right) - \left(M_2 G/s_2\right) \tag{2}$$

Now, the escape speed from two masses can be written

$$u_{1,2} \equiv \sqrt{-2U_{1,2}} \equiv \sqrt{-2(U_1 + U_2)} = \sqrt{(2M_1 G/s_1) + (2M_2 G/s_2)} \tag{3}$$

which is the value, in the considered point E_p of the (escape) velocity $u_{1,2}$ which has to be referred (at the considered time), to the point, we may call it Centre of potential (C_p), where $|U_{1,2}|$ has the max value. Then, as $\sqrt{2M_1 G/s_1} = u_1$ and $\sqrt{2M_2 G/s_2} = u_2$ we also get

$$u_{1,2}^2 = u_1^2 + u_2^2$$

(4)

therefore the escape speed due to all the n masses in space becomes

$$u = \sqrt{\sum u_n^2} = \sqrt{-2U} = \sqrt{\sum 2M_n G / s_n}$$

(5)

with $\sum M_n = M_u$ the universe mass, $U\left(= -\sum M_n G/s_n = -\frac{1}{2}u^2\right)$ the total gravitational potential in the considered point E_p, and where u (function of U in E_p) can be called as total escape speed (toward the infinity), while the escape velocity u is referred to the centre C_p. Indeed, any unitary massive particle during its path toward the infinity, has to comply with the CoE, U + K = 0, where $K = \frac{1}{2}u^2$ giving to this particle a speed u (which depends on the location of the source) and yielding, for all the masses, the total energy equal to zero [Compliance of light with above relation E = U + K, is shown on Appendix B].

We assume now the equality c = u, hereafter supported by the estimated mass of universe and also by a cosmological reason: in fact, if c > u the energy of light will be lost forever and furthermore the observable masses, following the always increasing mass of light going toward the infinity, will also tend to the infinity moving away from each other. On the contrary, if c < u, all the masses in space (having speed lower than u), will tend to a gravitational collapse, whereas for c = u, the mass of light, tending to the infinity in an unlimited time, will avoid the two said events (collapse or dispersion).

Now the mass of universe, by some authors, is estimated [4] -[6] to be $M_u \cong 10^{53}$ kg; the same order of magnitude is given through the number ($\cong 10^{22}$) of observable stars [7] [8], and since from Earth the distribution of the masses appears to be homogeneous and isotropic, under our assumption $U_{\ast} = 0$, we may assume their density as decreasing toward the infinity like a function $\rho = \rho_c e^{-as}$ with $\rho_c \cong 9.2 \times 10^{-27}$ kg/m³ the critical density [9]. So the mass of universe can be written

$$M_u = \int_0^\infty 4\pi s^2 \rho_c e^{-as} ds = 4\pi \rho_c \int_0^\infty s^2 e^{-as} ds = \frac{8\pi \rho_c}{a^3} \cong 10^{53} \text{ kg}$$

(6)

yielding

$$a = \left(8\pi \rho_c / M_u\right)^{\frac{1}{3}} \cong 1.3 \times 10^{-26} \text{ m}^{-1}$$

(7)

On Earth, the variation of potential due to an increase of the distance ds, can be written as dU = −dmG/s where dm = ρ4πs²ds with ρ = $\rho_c e^{-as}$, hence the potential on Earth becomes

$$U_0 = -\int_0^\infty \left(4\pi s^2/s\right)G\rho_c e^{-as}\,ds = -4\pi\rho_c G\int_0^\infty e^{-as}s\,ds = -4\pi\rho_c G/a^2 \cong -4.5\times10^{16}\ \text{J}$$

$$(8)$$

Now, according to (5), on Earth it is

$$u_o = \sqrt{-2U_o} \cong \sqrt{9\times10^{16}} \cong 3\times10^8\ \text{m/s}$$

$$(9)$$

Therefore, on Earth, $u_0 = c_0$, so that $\frac{1}{2}c_o^2 = -U_o$, and, in general we may argue

$$c = \sqrt{-2U} = u$$

$$(10)$$

The equality c = u, which implies the massiveness of light, means that, along any free path, the speed of light only depends on the value of the potential along that path.

[As for the relation $c_o^2 = -2U_o$, the Harvard tower experiment has shown that the fractional change in energy

(of light) is given by $\delta E/E = -gh/c^2$, and since the term gh is the variation of potential from the ground to the height h, we may guess that c^2 has to be related to the total gravitational potential, as also shown on §6].

ANNUAL VARIATION, ON EARTH, OF THE TOTAL ESCAPE SPEED

On Earth a small variation of the total escape speed u_o, from (9), can be written as

$$\Delta u = \Delta U/u_o$$

$$(11)$$

where ΔU is the variation of the total potential on Earth, mainly due to the variable distance Earth-Sun.

So considering the eccentricity e (=0.0167) of Earth's orbit around the Sun, between their average distance d (=1.5 ′ 10^{11} m) and their shortest distance (Perihelion) p = (1 – e)d, and with $u_0 = 3$ ′ 10^8 m×s^{-1}, the (11) gives

$$\Delta u_e = -\frac{\Delta U_S}{u_o} = \frac{\left[\left(\dfrac{M_S G}{p}\right)-\left(\dfrac{M_S G}{d}\right)\right]}{u_o} = +0.05\ \text{m}\cdot\text{s}^{-1}$$

$$(12)$$

with Δu_e the variation of u due to Earth's orbit eccentricity, ΔU_S the variation of potential on Earth due to Sun between the two said distances, with M_S the mass of Sun. Hence from Aphelion to Perihelion, one should find Δu_{AP} (=Δc_{AP}) = +0.10 m×s^{-1} and we note that this variation is compatible with the accuracy of the measured value of c = 299792458 m×s^{-1}. Due to Earth's rotation, there is also a daily variation which, from midnight to noon, is of the order of $\Delta u_r \cong 2 \times 10^{-4} \, m \cdot s^{-1}$; so, on Earth, u_o is practically constant during one year, as it is for the measurements of the speed of light.

INVARIANCE OF C FOR A PARTICULAR PARTICLE, HERE DEFINED, AND RELATED DOPPLER EFFECT

Here we show that the Galileo's velocities composition law, (related to point-particles), cannot be correctly applied to a particle, (hereafter called photon), defined as follows:

"Longitudinally-extended, elastic non divisible particle emitted at speed u by a source during an emission time T, and moving along one ray (continuous succession of photons), where two consecutive photons cannot be separated along a free path (constraint of non separation)".

Of course, more photons emitted during an emission time T need an equal number of rays.

Calling front and tail the extremities of a photon, the constraint of non separation implies that, along a ray, any tail corresponds to the front of the next photon.

Referring to Figure 1(a) (where E_p is the location of S at t = 0 and S_T its location at t = T), since the escape velocity c (=u) of an emitted photon (AB) is referred to the Centre of potential C_p, during its emission time ($0 \leq t \leq T$), the term $v_{CpA} = u$ should appear as the velocity of its front (A) from C_p.

The source S may have a velocity v_{CpS} from C_p, thus writing $v_{CpA} = v_{CpS} + v_{SA}$ we should find $v_{SA} = u - v_{CpS}$; this means that each photon emitted around the source should have a length $\lambda' = |v_{SA}T| = |(u - v_{CpS})|T$ depending on v_{CpS}, but this is contrary to the experience showing that if the source is fixed to its initial Emission point E_p (that is the point where S is located at the start of the emission) the emitted photons, referring to E_p, have equal characteristics. Thus, during the emission of a photon, the velocity of its front, (to comply with these equal characteristics), has to be referred to the initial Emission point E_p, therefore, see Figure 1(b), where E_p is our reference frame, as for the front A, for definition, we have

$$v_{EpA} = u \tag{13}$$

[This condition also allows the whole photon to have a velocity u referred to C_p, as shown on Figure 1(d)].

Now the velocity of the front A, with respect to S, from (13), becomes

$$\mathbf{v}_{SA} = \mathbf{v}_{SEp} + \mathbf{v}_{EpA} = \mathbf{u} - \mathbf{v}_{EpS} \tag{14}$$

and still referring to Figure 1(a), (where S_T is the location of S at t = T), should S be fixed to E_p (that is $v_{EpS} = 0$), the length λ of each photon, after the emission time T, from (14) becomes $\lambda = v_{SA}T = uT$, while, in general, it is

$$\lambda' = \mathbf{v}_{SA}T = \left(\mathbf{u} - \mathbf{v}_{EpS}\right)T = \lambda - \mathbf{v}_{EpS}T \tag{15}$$

where λ' is the photon AB emitted with the source in motion from E_p.

Referring now to Figure 1(c), if a generic Observer O is our Reference frame, we can write

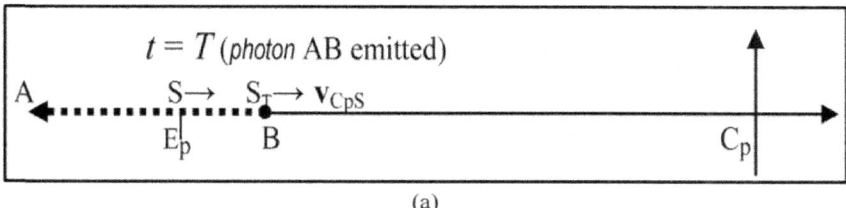

(a)

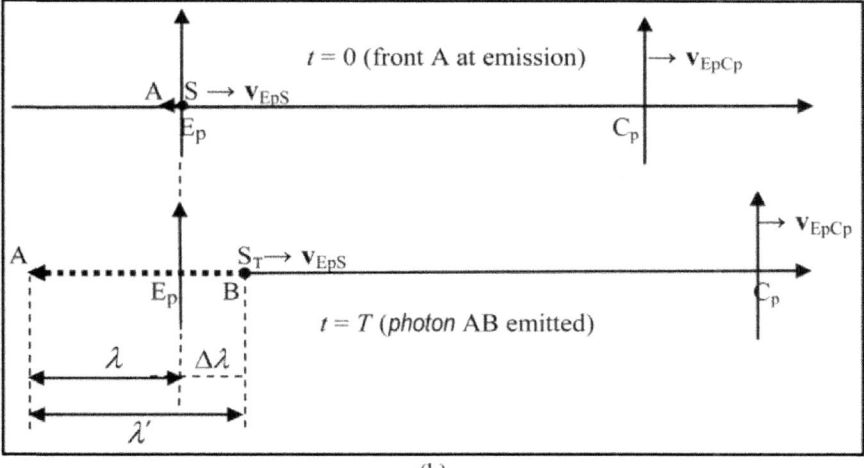

(b)

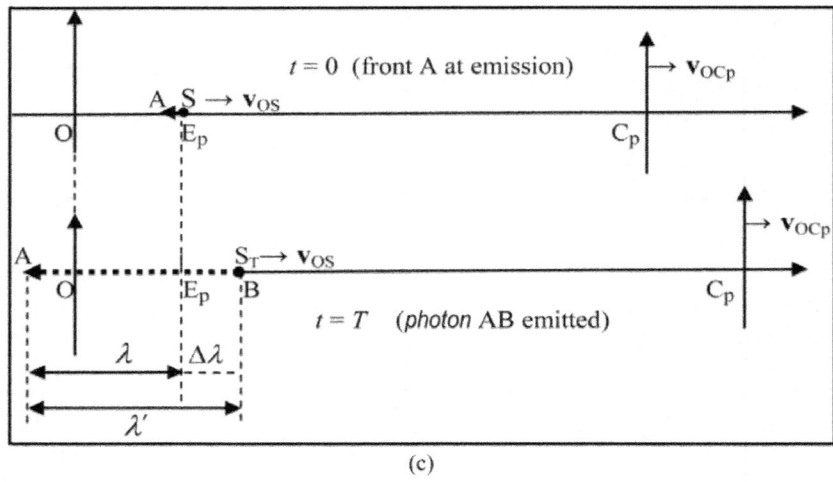

(c)

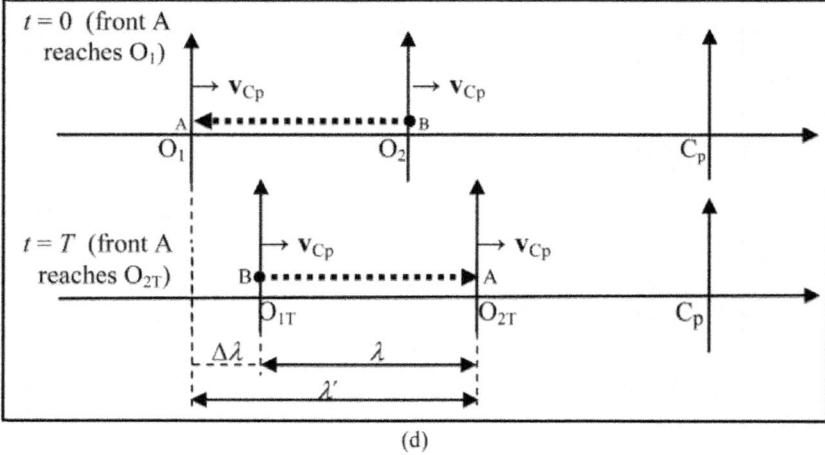

(d)

Figure 1: (a) Photon AB emitted under the supposed condition $v_{CA} = u$; (b) Emission of a photon AB referred to the initial Emission point E_p; (c) Emission of a photon AB referred to the generic Observer O; (d) Measurement of the speed of a photon (AB) reflected by O_1.

$$\lambda' = v_{SA}T = \left(v_{SO} + v_{OEp} + v_{EpA}\right)T = \left(v_{SEp} + u\right)T = \lambda - v_{EpS}T \tag{16}$$

where λ' is the photon emitted while the source is in motion, with velocity v_{OS}, from the Observer, and once more, if $v_{EpS} = 0$ (S fixed to E_p), we find $\lambda' = \lambda = uT$ (If S is now our Reference frame, and v_{EpS} is the velocity of S from E_p, we still have the (15)).

Thus, after the emission time T, as for a source receding from the front of the considered photon, as in Figure 1(b) (or Figure 1(c)), the length λ' (for any Observer) turns out to be

$$\lambda' = uT + vT = \lambda + \Delta\lambda = \lambda(1 + \beta)$$

(17)

where v $(=|v_{Eps}|)$ is the speed (referred to E_p) of the source S (along the direction E_pS), $\Delta\lambda$ $(=vT)$ is the path covered by S during T, and where $\beta = v/u$, and we point out that the length λ' may change, along a free path, and under constant potential, only during its emission.

Now, the speed of a point-particle is defined through two Observers, while the speed u of a photon, because of its variable length during its emission, does not correspond to the speed of any point of it, hence we must consider its length referred to the time T (transit time) the photon (front to tail)needs to cross one Observer, so it has to be defined

$$u' = |\lambda - vT|/T' = \lambda'/T'$$

(18)

[As for this definition, let us consider a system composed of two balls connected through an elastic thread and let them fall in vertical line: during the fall, each part of the system has different speed, so we define the speed of the whole system according to Equation (18)].

Returning now to Figure 1(c), for the Observer O, the transit time T' of the photon AB is given by the time the front A spend to cover the path λ, that is $T(\lambda/u)$, plus the time the tail B needs to cover the path $S_T - E_p = \Delta\lambda$; now, once the photon AB has been emitted (at t = T), the velocity of the front A has to be the same as any other part of the emitted photon, hence the time needed by B to cover the path $\Delta\lambda$ is $\Delta T = vT/u$, giving

$$T' = T + \frac{\Delta\lambda}{u} = T + \frac{vT}{u} = T(1 + \beta)$$

(19)

Now, according to (18), the speed of the photon AB, referred to O, becomes

$$u' = \frac{\lambda'}{T'} = \frac{\lambda(1 + \beta)}{T(1 + \beta)} = u$$

(20)

showing that the speed of photons emitted by a source S is invariant for any Observer, in spite of any speed of S with respect to the Observer [After the emission, each part of the photon has same velocity u, meaning that, during the emission, it is the velocity of its inner part to vary in order to change its length in the given time T].

As for an emitted photon, the measurement of c (through the method d/t) implies its absorption and reflection by an Observer. In this way, the Observer becomes the source of a new photon, with the Observer/Source located in the Emission point E_p, so we may refer to Figure 1(b), with the source fixed in E_p, finding u' = λ/T = u.

[Anyhow, we may obtain the same result ($u\rangle = \lambda'/T' = \lambda/T = u$) as follows: the measurement of c (through the method d/t) implies two Observers at a constant relative distance $O_1 O_2$; on these bases, see Figure 1(d) where C_p is now our Reference frame, after the reflection of the photon from O_1, at t = T, the path covered by the front A to reach O_{2T} is given by $O_1 O_{2T}$ that is $\lambda\cent = \lambda + \Delta\lambda$ where λ is the length of the emitted photon AB and where $\Delta\lambda = v_{Cp}T/c$, with v_{Cp} the speed of our frame $O_1 O_2$ with respect to C_p, yielding $\lambda' = \lambda(1 + \beta)$ where $\beta = v_{Cp}/c$. The time needed by the front A to cover the distance $O_1 O_{2T}$ is T' = T + $\Delta\lambda/c$ = T(1 + β), thus the measured speed (referred to the two Observers) becomes $c\rangle = \lambda\rangle/T\rangle = \lambda/T = c$ in spite of any velocity of the co-moving Observers $O_1 O_2$ with respect to C_p (Anyhow, the Observer O_1 could state, for the front A, a velocity $v_{O_1 A}$ different from u, if he could measure such a speed)].

For any Observer, the frequency of photons of the same ray has to be defined as n' = n/t with n the number of photons crossing the Observer during a time t; for t = T' (transit time of one photon), it is n = 1, thus n' = 1/T', so from (19) we get

$$\nu' = \frac{1}{T'} = \nu/(1+\beta) \tag{21}$$

showing that for v = 0, that is β = 0, we have $n\rangle$ = n, which is also valid if the Observer (O) and the source (S) belongs to different potential: in fact, for O and S at reciprocal rest, the number of photons emitted by S in a unit time has to be equal, in the same time, to the number of them crossing O (like, for instance, the number of balls falling from the top of a tower with respect to an Observer at the tower base), and this implies $n_s = n_o$.

Now, the Figure 1(c), where a source emits a photon while it is in motion from the Observer O, also represents a longitudinal Doppler effect, which, in general, can be written a

$$\lambda' = \lambda(1\pm\beta); \quad \nu' = \nu/(1\pm\beta) \tag{22}$$

with the sign + for S receding from the Observer, while the sign − is for S approaching it.

Hereafter we get our equations regarding both the generic and the transverse Doppler effect, followed by our relations regarding a source (of light) circling around an Observer.

To get a general relation for the Doppler effect, let us consider, see Figure 2(b), referring to the Observer O, a source S, located in E_p (at t < 0), at rest with O. During this time let S emit photons having length λ (=uT) and let E_pO = λ. Then, at t = 0, let S start to move from E_p toward S_T(reached at t = T), with velocity v (referred to O) along the generic direction a-a. Now, during the path E_pS_T, let S emit a photon λ¢ toward O. (On Figure 2(a), the small arrow inside the triangle E_pOS_T represents the partial λ¢ during its emission.) At t = T (end of emission), according to (16) we have λ¢ = λ – vT, thus the length of λ¢, assuming v = u, so to consider E_pO = NO, with $E_pN \perp S_TO$, becomes

$$\lambda' = \lambda + vT\cos\alpha = \lambda\left(1 + \beta\cos\alpha\right) \tag{23}$$

As for the transit time T', as before, we can write $T' = T + \Delta\lambda/u = T + \left(vT\cos\alpha\right)/u$

.

which can also be obtained considering that the front of λ¢, following the tail of λ (thus directed toward O), takes a time T to reach O from E_p, while the tail of λ¢, emitted in S_T, has to cover the path $S_TO = S_TN + NO$, spending the time ΔT = (vTcosα)/u for the path S_TN, plus the time T for the path NO (equal to E_pO for v = u), giving

$$T' = T + \left(\frac{vT}{u}\right)\cos\alpha = T\left(1 + \beta\cos\alpha\right) \tag{24}$$

yielding

$$v' = v/\left(1 + \beta\cos\alpha\right) \tag{25}$$

thus $u' = \lambda'/T' = \lambda/T = u$ (For an opposite direction of S we get λ' = λ(1 – βcosα) and T› = T(1 – βcosα).

[Ray S_{2T}FO: referring now to Figure 2(b), if S, between T and 2T, is still moving with same velocity v, the emitted photon $\boldsymbol{\lambda''}$ will have same length as $\boldsymbol{\lambda'}$, thus its front (at t = 2T) will reach a point F, (corresponding at the same time to the tail of $\boldsymbol{\lambda'}$), at a distance $FO = S_{2T}N = vT\cos\alpha$ from O, so the ray Source-Observer, at t = 2T, becomes the line S_{2T}FO].

Transverse Doppler effect: referring to Figure 3(a) left part, where $S_TO \perp E_pS_T$, should S start to move at t = 0 from E_p to S_T while emitting the photon λ', according to (16) we can write $\lambda' = \sqrt{\lambda^2 - (vT)^2} = \lambda\sqrt{1 - \beta^2}$ (valid for S approaching O) (26)

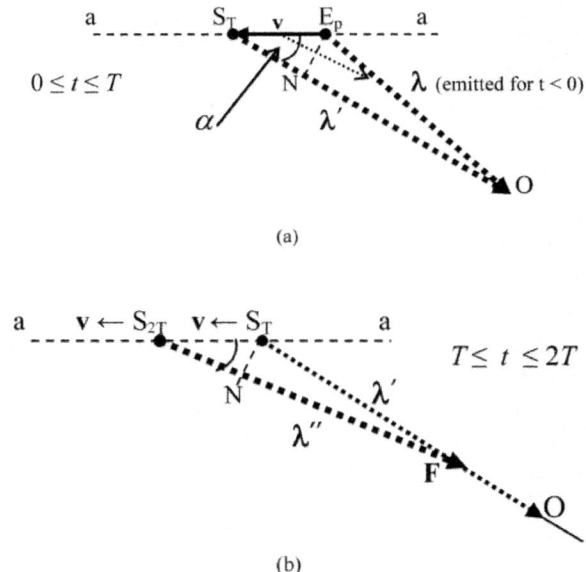

(a)

(b)

Figure 2: Doppler effect for a photon, general case.

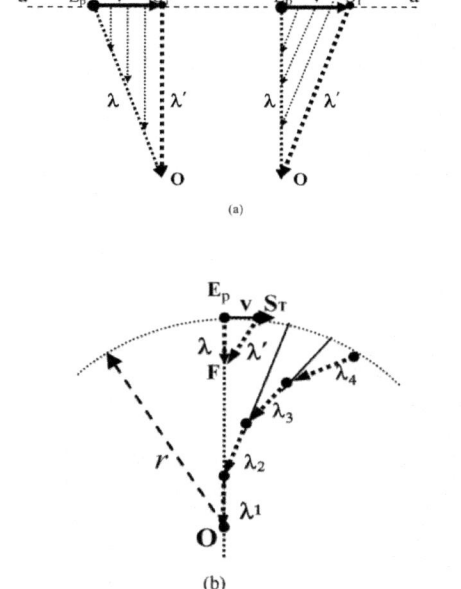

Figure 3: (a) Transverse Doppler effect; (b) Source circling around the Observer O.

where λ' is the length corresponding to S_TO, while λ corresponds to E_pO. On the contrary, seeFigure 3(a) right part (where the source is receding from the Observer), it will be $\lambda' = \sqrt{\lambda^2 + (vT)^2} = \lambda\sqrt{1+\beta^2}$ (valid for S receding from O) (27)

Then $T' = T + \dfrac{\Delta\lambda}{u} = T + \dfrac{\lambda'-\lambda}{u} = T + (\lambda'/u) - \dfrac{\lambda}{u} = \lambda'/u = \lambda\sqrt{1+\beta^2}/u$, hence $c' = c$.

Regarding a source circling around an Observer O, on Figure 3(b) the line E_pO represents a succession of photons λ already emitted when S is fixed in E_p, while E_pF represents the last of them (or it could represent the last photon emitted by S when reaching E_p). Then, at $t = 0$ let S start to move from E_p with velocity v toward S_T.

Now, because of the constraint of non separation, the front of the first photon λ' emitted when S is moving between E_p and S_T, has to reach, in F, the tail of previous photon, so, according to (16) the length of every photon λ' (emitted while S is moving along the orbit r) will be

$$\lambda' = \sqrt{\lambda^2 + (vT)^2} = \lambda\sqrt{1+\beta^2} = \lambda\sqrt{1+\omega^2 r^2/c^2} \qquad (28)$$

thus

$$T' = T\sqrt{1+\beta^2} = T\sqrt{1+\omega^2 r^2/c^2} \qquad (29)$$

with r the orbit radius, ω the angular speed, giving to any whole photon the speed $c\rangle = c$.

Figure 3(b) also shows a path $(\lambda_1$-$\lambda_4)$ of a ray directed toward O (the lines connecting the photons λ_2 and λ_3 to the orbit give the point where the source is located at the end of their emission).

PHYSICAL CHARACTERIZATION OF THESE PHOTONS

Now, similarly to a fluid flowing in a pipe (whose kinetic energy is $K = \frac{1}{2}mv^2$ with m the mass passing in 1 s)the kinetic energy of light flowing along one ray (according to our definition, photons are also massive), has to be expressed with $K_c = \frac{1}{2}mc^2$ with m the mass of the particles passing in 1 s along one ray. Anyhow, the total energy of light flowing along one ray is $E = mc^2$ as also proved by the evidences of nuclear reactions like $n + p ® d + \gamma$: indeed, in this reaction [10] , the lost mass, known through mass spectrometers, corresponds to the value $m = E/c^2$ where E $(=hc/\lambda)$, (as λ is measured), is also known, so $E = mc^2$ represents the total energy of light flowing along one ray (λ_{meas} is obtained [11] through the value $\lambda_{meas}/(d_{220})$ given at pag. 369, where (d_{220}) is given at page 410).

So, writing $E = \frac{1}{2}mc^2 + \frac{1}{2}mc^2$ we may infer that each of these particles is provided with an internal energy ($K_i = \frac{1}{2}mc^2$) equal to its kinetic energy. Now, equating mc^2 to $h\nu$ we get

$$mc^2 = h\nu(= E) \tag{30}$$

where m written as

$$m = \frac{h\nu}{c^2} \equiv \gamma\nu \; (kg) \tag{31}$$

is the mass of light, with frequency $\boldsymbol{\nu}$, passing along one ray in 1 s, while the constant

$$\gamma \equiv h/c^2 \left(= m/\nu\right) = 7.372495 \times 10^{-51} \; kg \cdot s \tag{32}$$

is the mass of light passing along one ray during T, we may call it "mass of one photon"; so one finds

$$h = \gamma c^2 = mc^2 T \tag{33}$$

and therefore the Planck's constant represents the energy of one photon. The energy of these particles passing in 1 s along one ray (energy of one ray of light) can now be written as

$$E = K_c + K_i = \frac{1}{2}mc^2 + \frac{1}{2}mc^2 = mc^2 = h\nu = \gamma c^2 \nu \tag{34}$$

On the above bases, the total energy of light emitted by a source is given by $n_r mc^2$ with n_r the number of rays, and since m is the mass of light passing along one ray in 1 s, this unitary (for unit of time) energy shall be equal to the supplied power P during 1 s, thus $n_r mc^2 = P$, hence the total mass lost per second m_T ($°n_r m$) by a source of light becomes

$$m_T = n_r m = P/c^2 \tag{35}$$

So, for a 1 W lamp, we get $m_T = P/c^2$ @ $1.1 \; ' \; 10^{-17} \; kg \times s^{-1}$, while the number n_r of rays is

$$n_r \left(= m_T/\gamma\nu\right) = P/c^2\gamma\nu = P/h\nu \tag{36}$$

in our case, n_r @ $3 \; ' \; 10^{18}$ rays. We point out that for a given power P, the higher is the frequency, the lower is the number of rays, as shown by (36) written as $n_r\nu$ = P/h. The number of photons emitted in 1 s becomes:

$$n_\gamma \left(= n_t v\right) = Pv/hv = P/h \tag{37}$$

which, for P = 1 W, gives $n_\gamma = h^{-1}$ (=1.5 ′ 10^{33} photons/s), so the inverse of Planck constant corresponds to the number of photons emitted in 1 s by a source of unitary power (This great number of photons (having emission time T at speed c) can be regarded as a wave function).

Now the momentum of the photons passing along one ray in 1s, considering their kinetic energy only, that is $K_c = \frac{1}{2} mc^2$, according to Newtonian mechanics should be written as

$$\mathbf{p} = m\mathbf{c} = \gamma v\mathbf{c} = \gamma \mathbf{c}/T \tag{38}$$

but considering both their kinetic and their internal energy, that is $E = mc^2$ we obtain

$$\mathbf{p} = 2m\mathbf{c} = 2\gamma v\mathbf{c} = 2\gamma \mathbf{c}/T = \left(\gamma \mathbf{c}/T\right) + \left(\gamma \mathbf{c}/T\right) \tag{39}$$

REVISITATION OF THE HARVARD TOWER EXPERIMENT AND TIME DILATION

Referring to Harvard tower experiment [1]-[3], simply represented on Figure 4, where h is the tower height, calling c_0 the value of c on Earth's surface at the tower base and c_h its value on its top, the variation $c_h - c_o$) from the tower base to its top, from (11), for c = u, becomes

$$\Delta c = c_h - c_o = -\Delta U/c_o \tag{40}$$

where $\Delta U = \left(U_{Eh} - U_{Eo}\right)$ is the variation of the total gravitational potential U, due to Earth, from the tower base to its top. As $U_{Eo} = -M_E G/r_o$ and $U_{Eh} = -M_E G/r_h$ where M_E is the Earth's mass and r its radius, we get $\Delta U = M_E Gh/r^2$ where h ($=r_h - r_o$) is the tower height, yielding

$$\frac{\Delta c}{c_o} = \frac{c_h - c_o}{c_o} = -\frac{M_E Gh}{r^2 c_o^2} = -gh/c_o^2 \tag{41}$$

showing that, on the top, where $|U_h| < |U_o|$, it is $c_h < c_o$, with $c_h = c_o\left(1 - gh/c_o^2\right)$.

Now, let S be a Mossbauer source and A an appropriate absorber; if they are close to each other (for instance, at the tower base), the absorber is in resonance with the source.

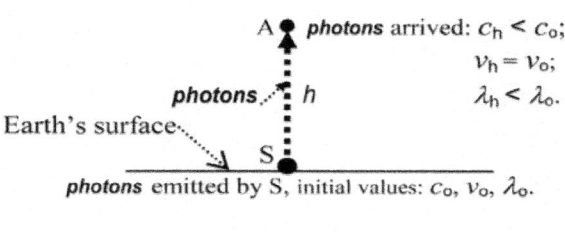

(a)

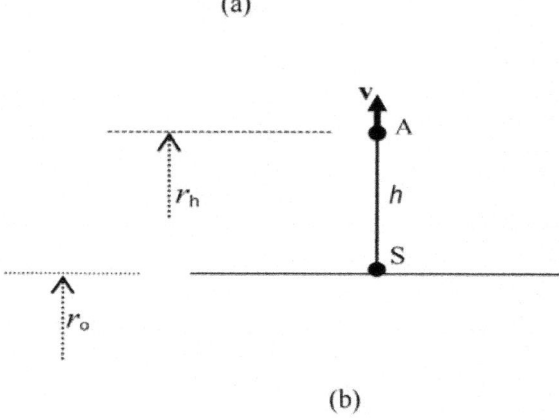

(b)

Figure 4: Harvard tower experiment scheme, with the source at the base. (a) S and A at rest at a different level h. In A, $\lambda_h < \lambda_o$, so the detector observes a gravitational blue-shift; (b) Source and absorber relative motion (v) to compensate the gravitational blue-shift through the Doppler effect.

Then, see Figure 4(a), with S at the tower base and taking A to its top, while S and A are at rest, the frequency of the emitted photons (i.e. the number of photons emitted along the direction SA per unit of time) has to be equal to the photons received by A, that is $n_h = n_o$ and since $c_h < c_o$, it must be $\lambda_h < \lambda_o$ (indeed $\Delta\lambda/\lambda_o = \Delta c/c_o$), so, contrary to ToR, a blue-shift effect for A.

[On Figure 4(a) (photons arrived to the top), according to $n_h = n_o$, it seems to be $E_h/E_o = hn_h/hn_o = 1$ (here h is the Planck constant), but the (33) shows that $h = \gamma c^2$ with γ (representing the mass of light passing during T along one ray), an effective constant, so that we get $E_h/E_o = (c_h/c_o)^2$ which shows a decrease of the energy of light from S to A].

Indeed, with S on the base emitting toward A on top, A goes out of resonance and since on our bases $n_h = n_o$, the non-resonances physically related to a variation of λ, whereas in the Harvard tower experiment [3] , "no mention has been made of frequency or wavelength".

Thus, to restore the resonance through the Doppler effect (i.e. to increase the photon length from its value λ_h in A to its initial value λ_o in S), since $\lambda_h < \lambda_o$, A and S, see Figure 4(b), have to recede from each other with speed v complying with (17), here written $\lambda_o = \lambda_h + vT$, giving $(\lambda_h - \lambda_o)/\lambda_o = -vT/\lambda_o$.

Therefore, since $\Delta\lambda/\lambda_o = \Delta c/c_o$ (as $n_o = n_h$), we find $\Delta c/c_o = -vT/\lambda_o$ and comparing to (41) we get $-vT/\lambda_o = -gh/c_o^2$, so the relative speed between S and A becomes

$$v = gh/c_o = 7.5 \times 10^{-7} \text{ m/s} \quad (\text{for } h = 22.5 \text{ m}) \tag{42}$$

[This value is also predicted by General Relativity (GR) which, implying a decrease of v for light moving from the base to the top, predicts an opposite direction of v with respect to the one shown on Figure 4(b); at this regard, Pound-Rebka [3] operated in order to determine (through the value of v, obtained moving the source sinusoidally) the variation of energy of a beam on the upward and downward path, without any indication (because of the low value of v), about the direction of the compensating speed].

Now, if we take S to the tower top, with A located on its base (see Figure 5 which is referred on our bases), the experiment shows that the absorber goes out of resonance.

Now, according to Relativity, taking S to the top, the initial frequency of the light should be $n_h = n_o$, which, on our bases, is wrong: with S on the top, see Figure 5(a), the (10) written as $c^2 + 2U = 0$, between top and base gives $c_h^2 + 2U_h = c_o^2 + 2U_o$, where $U_h - U_o = gh$, giving $c_o^2 = c_h^2 + 2gh$, then $c_o = (1 + 2gh/c_h^2)^{1/2}$, and since 2gh

$= c^2$ we can write $c_o = c_h(1 + gh/c^2)$, that is $c_h < c_o$ as showed by (41), but what about n_h and λ_h?

Well, referring to previous Figure 4(a), with source S on the base, the length of photons arriving to the top varies from λ_o to λ_h (with $\lambda_h < \lambda_o$), therefore if S has been taken now to the top, should their initial length be λ_h, at their arrival to the base, their length should be λ_o, and since the resonance, as seen, depends on λ, the Absorber A (on the base, see Figure 5(a)), should be now in resonance. Thus we can argue that taking the source on top, the photons initial length has to be $\lambda_h (= \lambda_o)$; then, as $c_h < c_o$ as shown by (41), it must be $n_h < n_o$, and in particular, according to (41) we get $\Delta c/c_o = \Delta v/v_o = -gh/c_o^2$, giving $v_o = v_h(1 + gh/c_o^2)$ which is the same as the one predicted by GR, but on our bases, this variation is due to a different initial frequency (as the source is now on the top), whereas for GR this variation is related to the path of the emitted light between two different levels. (Indeed on our bases the frequency remains constant during any path, should source and observer be at reciprocal rest).

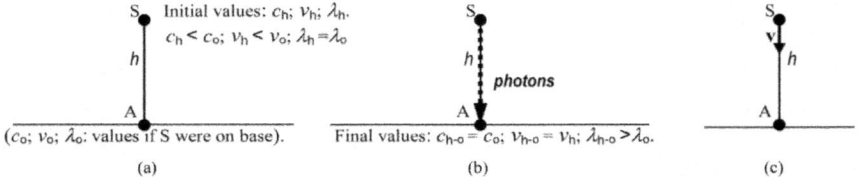

(a)   (b)   (c)

Figure 5: Harvard tower experiment scheme, with the source on the top. (a) S on top, S and A at rest: c_h, n_h, λ_h are the photons initial parameters on the tower top; (b) S and A at rest. When photons reach the base, $\lambda_{h\text{-}o} > \lambda_o$, so A observes a g-redshift; (c) S and A relative motion to compensate the g-redshift via Doppler effect.

Thus, see Figure 5(b), with S emitting from the top, S and A at rest, when the photons reach the base, as their final frequency ($n_{h\text{-}o}$) will remain the same as the initial one (that is $n_{h\text{-}o} = n_h$) and since $c_o > c_h$, it turns out that, along the path SA, λ will increase, and its variation, opposite to the one given by (41), yields now

$$\Delta\lambda/\lambda_o = gh/c_o^2 \text{ , giving } \lambda_{h\text{-}o} = \lambda_o\left(1 + gh/c_o^2\right).$$

Now, as $\lambda_{h\text{-}o} > \lambda_o$, the absorber, on the base, will observe a gravitational redshift so, to compensate it via Doppler shift, see Figure 5(c), S and A have now to move relative to each other in order to decrease the final length $\lambda_{h\text{-}o}$ to the resonance value λ_o; on the contrary, according to ToR, A and S should recede from each other.

[Still referring to Figure 5, taking S to the tower top, we have $c_h < c_o$ and $n_h < n_o$ implying, contrary to GR, a decrease of the energy of light to be emitted by S, $\left(E_h = h\nu_h = \gamma c_h^2 \nu_h\right)$, see (34), and therefore when these photons reach the base (where $c_{h\text{-}o} = c_o$) their energy becomes $E_{h\text{-}o} = \gamma c_o^2 \nu_h$ giving a reason to the loss of energy as for light arriving to Earth coming from sources located in points where $|U_S| < |U_o|$, and since, as seen, $\lambda_{h\text{-}o} = \lambda_o\left(1 + gh/c_o^2\right)$, we also give a cosmological reason to the high redshift of sources where $|U_S| = |U_o|$.

Time Dilation

Well, the experience shows that, on board of GPS satellites, the atomic clocks run faster by about 38 μs/day than the ones on ground, meaning that, in altitude, their ticking time, (or interval time, intending the minimum time counted), is shorter than the one on ground.

Now, the ticking time t of atomic clocks is proportional to their frequency, so on ground we can write $t_o = kv_o$ while in altitude $t_h = kn_h$ yielding $\Delta n/n_o = \Delta t/t_o$ where $\Delta t \; (=t_h - t_o)$ is the ticking time variation from ground to height h, with $\Delta n \; (=n_h - n_o)$ representing their frequency variation due to the gravitational potential variation.

Now, taking the sources (clocks) from ground to height h, the length of their photons, at emission, remains constant, $(\lambda_h = \lambda_o)$, thus, because of the variation of c (from ground to height h), it has to correspond an equal variation of n, so that the (40) can be written as

$$\Delta c/c_o = \Delta v/v_o = \Delta t/t_o = -\Delta U_E/c_o^2 \tag{43}$$

Now, GPS satellites have an orbit of r_h 26,600 km, that is an altitude h 20,200 km, as r_o 6400 km is the Earth's radius. Hence, the (43), because of the variation of the potential, the variation of the counted time during one day (Δt_{1d}), since in one day $t_{1d} = 86,400$ s, gives

$$\Delta t_{1d} = -\left(\frac{\Delta U_E}{c_o^2}\right) t_{1d} = -\left[\frac{M_E G(r_h - r_o)}{r_h r_o c_o^2}\right] 86,400 = -45.6 \; \mu s \tag{44}$$

where the sign means that the ticking time is decreasing, inducing the clocks to run faster. Then we have to take into account that the parameters of the photons emitted by atomic clocks on board of GPS satellites are changing because they are circling around the Earth.

Therefore, according to (29), that is $T' = T(1 + \beta^2)^{1/2}$ where T' is the time a photon needs to cross the Observer, during one day (86,400 s), since the orbital speed corresponds to two orbits every day (giving $v = 2(2\pi r_h/86,400) = 3870$ m×s^{-1}, and considering that for v = c we can write $(1 + \beta^2)^{1/2} @ 1 + \beta^2/2$, we get

$$\Delta t'_{1d} = (T' - T)86,400 = (\beta^2/2)86,400 = 7.2 \; \mu s \tag{45}$$

representing the variation of the counted time in one day due to the orbital speed of GPS satellite, and since this variation is positive, it has to be deducted from the negative one due to the potential variation, thus the total variation of the counted time on GPS satellites, in one day, becomes

$$\Delta t_{day} = \Delta t_{1d} + \Delta t'_{1d} = -38.4 \; \mu s/day \tag{46}$$

as observed. This equality also confirms that $\lambda_h = \lambda_o$ as for sources in altitude.

RED SHIFT

According to the Relativity, the gravitational red shift of light coming from the Sun, with M_S and R_S its mass and radius, is $z_S \cong M_S G/R_S c^2 = 2 \times 10^{-6}$. Now, for $s \lessapprox 20$ Mpc, with s the distance Earth-source, the observed shifts are in the range $z \cong \pm 10^{-2}$ [12] , while from $\cong 20$ to $\cong 40$ Mpc they tend to become always positive, and between $\cong 40$ and $\cong 900$ Mpc ($\cong 3$ Bly) the red shifts (here in the range $\cong 0.01 - 0.20$), practically follow the empirical Hubble's law z = H$_0$s/c; hence, since the value of the gravitational red shifts of a typical galaxy, intended to be $z_g \cong M_g G / R_g c^2$, should be of the order of 10^{-9}, the Doppler effect appears to be (as for the Relativity), the only satisfactory way to explain the observed blue shifts and also the high (cosmological) red shifts.

On the contrary, on our basis, disregarding any motion between a source and an Observer on Earth, which implies (as showed on §4) n = n$_0$, we get $c/\lambda = c_o/\lambda_o$, where n$_0$, c$_0$ and λ_o are observed on Earth, showing that for c$_0$ > c, it has to be $\lambda_o > \lambda$. Hence, the blue/red shifts observed on Earth can be expressed as

$$z \equiv \Delta \lambda / \lambda = \Delta c / c = (c_o - c)/c = (c_o/c) - 1 = \sqrt{U_o / U_s} - 1 \tag{47}$$

where $U_o \left(= -\frac{1}{2}c_o^2\right)$ is the potential on Earth, while U$_s$ the one on the source (at distance s). Thus the shift of a far source, disregarding the motion source-Earth, turns out to be the variation of c (as well as λ) during the path of light toward a different potential; for instance, going from Earth to Sun, and considering that along this path the main variation of potential ($\Delta U_{(s)}$) is due to the Sun only, we can write U$_S$ = U$_o$ + $\Delta U_{(s)}$ where

$$\Delta U_{(s)} = -(M_s G / R_s) + (M_s G / d) \cong -M_s G / R_s \tag{48}$$

With d the distance Earth-Sun; so on the Sun, $c_s = \sqrt{-2U_s} = \sqrt{-2(U_o + \Delta U_{(s)})}$ and then giving $\Delta c / c = z \cong 2.1 \times 10^{-6}$, while on the opposite path, Sun to Earth, it is $z \cong 2.1 \times 10^{-6}$, hence a blueshift (contrary to a red shift of the same value predicted by the Relativity).

$$c_s = \sqrt{c_o^2 + 2M_s G / R_s} \cong c_o \left(1 + \frac{M_s G}{R_s c_o^2}\right) = c_o + 635 \text{ m} \cdot \text{s}^{-1} \tag{49}$$

As for $s \lessapprox 40$ Mpc, according to (47), if U$_s$ (potential on the source) is, in absolute value, higher than the potential on Earth U$_o$, we get, on Earth, z < 0 (blue shift), and vice versa for $|U_s| < |U_o|$, hence, apart Doppler effects, these red/blue shifts indicate that the potential, from Earth to the sources in this space, may increase or decrease (and since for $s \gtrapprox 40$ Mpc, z is positive, we may also argue that our galaxy is close to the middle of the masses of universe); then, over this distance, it turns out that, on the related sources, U$_s$ is (in absolute value), always lower than U$_o$, and also tending to zero for z → ∞.

In the range $\cong 0.01 < z \lessapprox 0.20$ (where z follows the Hubble's law), the (47), written as

$$U_s = U_0 \big/ (1+z)^2$$
(50)

for z = 1 gives $U_s \cong U_o/(1+2z)$ yielding (through a simple artifice) to $U_s \cong U_o(1-2z)$ (valid for z = 1) (51) which shows that, for z = 1, U_s depends linearly on z; in particular, Table 1 shows that, in the said range ($\cong 0.01 - 0.20$), the values given by (51), are practically the same as the ones given by (50).Table 1 also shows the values of U and c for various values of z.

Table 1: Calculated values of U and c related to the observed redshifts on Earth. The 4th column is referred to Equation (50); the 5th to (51); the 6th to (47)

Redshift z	$s = zc/H_o$ Mpc	1 Mpc 3.3×10^6 ly light years	$U_s/U_o = 1/(z+1)^2$	$U_s/U_o = (1-2z)$	$c_s/c_o = 1/(z+1)$
<|0.01|	<43	<140 Mly	0.98 - 1.02	0.98 - 1.02	0.99 - 1.01
0.01	43	140 Mly	0.98	0.98	0.99
0.05	215	700 Mly	0.90	0.90	0.95
0.20	860	2.8 Bly	0.69	0.60	0.83
1.0			0.25		0.50
5.0			0.028		0.17
9.0			0.010		0.10

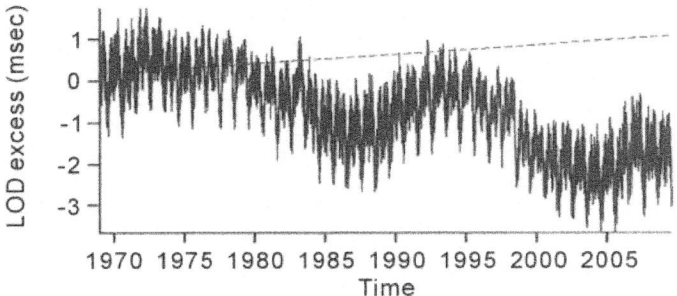

Figure 6: Length of day since 1969, when Earth-Moon laser ranging data started to be collected. The dotted line shows the 2.3 msec·cy^{-1} trend expected as a consequence of momentum conservation, when it is assumed that, using laser ranging, an actual increase of Earth-Moon distance is measured. LOD data comes from the EOP 05. C04 series (Bizouard and Gambis 2009), as provided by the Earth Orientation Centre (http://hpiers.obspm.fr). The Figure has been taken from the paper of Sanejouand, Y.H., Empirical evidences in favour of avarying-speed-of-light, arXiv 0908.0249.

CONCLUSIONS

We showed that, on our basis, c corresponds to the total escape speed u which is practically constant on Earth (see §3), while the annual variations of c, due to the eccentricity of the Earth's orbit, are well shown on following Figure 6.

In fact, when the Earth (on perihelion) approaches the Sun, because of the distortion of the shape of Earth, the angular speed of Earth should decrease, hence, the length of day (LOD)should increase; moreover, during the approach Earth-Sun, due to the increase of the absolute value (on Earth) of U, the speed of light (on Earth) has also to increase, thus the LOD and c should show, at the same time, annual peaks, as shown on Figure 1, which may also represent, in another scale, the annual variations of the speed of light, since 1969.

Now we point out our attention on Compton effect: indeed, through the relativistic Doppler effect equations, one cannot get the Compton equation (which can be found in other ways), whereas through our Doppler effect Equation (17), we can obtain it; well, one can observe that the Compton effect is not a Doppler effect, but in this case why we get the Compton equation? Is it a coincidence or the relativistic Doppler effect equations are not correct?

Finally, regarding the Harvard tower experiment, the Relativity, as for the compensating speed source-Observer (in order to restore the resonance between them), predicts opposite directions with respect to the ones we have obtained: we hope that now (after 50 years from the related experiment) an appropriate (similar) experiment will give a sure answer.

APPENDIX

Appendix A (Compton Effect)

Here, see Figure A1, an incident photon (length λ, frequency n), ejects a circling electron (m_e) but there is also a reflected photon (length λ' frequency ν') so the electron, while emitting a photon λ' toward the Observer A, represents a source in motion from A along the direction w, thus (considering the component w_A of w) there is an indubitable Doppler effect.

Now, on the basis that the scattered photon starts to be reflected at the same time when the incident photon starts to hit the electron, and since T' ($=1/\nu'$) is the emission time of the photon ν' , it turns out that T' is also the whole interaction time, meaning that there is not a complete absorption of the incident photon followed by an emission: this means that the internal energy of the photon is not involved in this action, hence the momentum transferred from the incident light to the electron is p = mc ($=\gamma c/T$) as per (38), and the same value p = mc is the momentum transferred from the scattered photon to the electron.

Therefore, the Conservation of Momentum (CoM) along the direction normal to w, becomes $(\gamma c/T')\sin\theta = (\gamma c/T')\sin\theta'$ giving $\theta = \theta'$.

Moreover, the length of the reflected photon, for the Observer, according to (17) is

$$\lambda' = \lambda + \Delta\lambda = \lambda(1+\beta)$$

(52)

where $\Delta\lambda = w_A T'$ and where $w_A = w\cos\theta$ is the component of the electron speed along the direction of the Observer A and T' $(=1/\nu')$ is, for A, the photon transit time, so we get

$$\lambda' - \lambda (\cong \Delta\lambda) = wT'\cos\theta.$$

(53)

Now the CoM along w is $(\gamma c/T')\cos\theta + (\gamma c/T')\cos\theta = m_e w$ giving

$$wT' = (2\gamma c\cos\theta)/m_e$$

(54)

and plugging this value into (53) we get

$$\lambda' - \lambda (\cong \Delta\lambda) = 2\gamma c\cos^2\theta/m_e = 2h\cos^2\theta/cm_e.$$

(55)

Now, $2\theta + \varphi = \pi, \Rightarrow \theta = (\pi - \varphi)/2$, hence $\cos\theta = \sin\varphi/2$ and therefore

$$\Delta\lambda = 2h\sin^2(\varphi/2)/cm_e$$

(56)

and since $2\sin^2(\varphi/2) = (1-\cos\varphi)$, we get the Compton equation:

$$\Delta\lambda = \lambda' - \lambda = h(1-\cos\varphi)/m_e c,$$

(57)

which cannot be obtained through the relativistic Doppler effect equation which, as for a source receding from the Observer, is $\lambda' = \lambda(1-\beta^2)^{1/2}/(1-\beta)$.

φ: angle between the direction of the
    incident **photon** and the scattered one (λ')

θ : angle between the direction of the
    incident **photon** and the recoiled electron

θ': $(= \pi - \varphi - \theta)$: it is shown that $\theta' = \theta$.

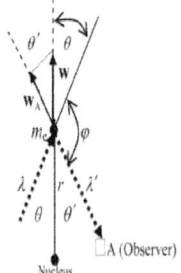

Figure A1: Compton effect.

Appendix B

Regarding the Conservation of Energy (CoE), E = U + K = 0, leading to $U = -\frac{1}{2}u^2$ the term K is the unitary (per unit of mass) kinetic energy of a massive particle, hence as for the light (having total energy, see (34), $E = K_c + K_i = \frac{1}{2}mc^2 + \frac{1}{2}mc^2$ with K_i its internal energy), we have to consider, to comply the CoE, its kinetic energy only, giving $U = -\frac{1}{2}c^2$, like any other massive particle. Anyhow, one can observe that the internal energy of light ($K_i = \frac{1}{2}mc^2$) of photons going toward the infinity could be lost, but this is not the case: indeed, we have seen on §6 (see also Figure 4(a)) that the frequency of a photon is constant along its path, and since, see (34), $K_i = \frac{1}{2}mc^2 = \frac{1}{2}\gamma vc^2 = \frac{1}{2}\gamma \lambda^2 v^3$, it turns out that toward the infinity, where $c_\infty \to 0$, we get $\lambda_\infty \to 0$, yielding $K_{i\infty} \to 0$ (In other words, the internal energy of light only depends on the length of their photons).

REFERENCES

1.   Pound, R.V. and Rebka Jr., G.A. (1960) Physical Review Letters, 4, 337. http://dx.doi.org/10.1103/PhysRevLett.4.337

2.   Pound, R.V. and Snider, J.L. (1964) Physical Review Letters, 13, 539. http://dx.doi.org/10.1103/PhysRevLett.13.539

3.   Pound, R.V. and Snider, J.L. (1965) Physical Review, 140, B788.http://dx.doi.org/10.1103/PhysRev.140.B788

4.   Kragh, H. (1996) Cosmology and Controversy. Princeton University Press, Princeton, 212.

5.   Gogberashvili, M., et al. (2012) Cosmological Parameter. ArXiv:physics.gen-ph, 2.

6.   Immerman, N. (2001) Nat'l Solar Observatory: The Universe. University of Massachusetts, Sunspot.

7.   Gott III, J.R., et al. (2005) The Astrophysics Journal, 624, 463-484.

8.   Van Dokkum, P. (2010) Nature, 468, 940-942. http://dx.doi.org/10.1038/nature09578

9.   Schutz, B.F. (2003) Gravity from the Ground Up. Cambridge University Press, Cambridge, 361.http://dx.doi.org/10.1017/CBO9780511807800

10.   Halliday, D. and Resnic, R. (1981) Fundamental of Physics. John Wiley & Sons, New York.

11.   Mohr, P.J. and Taylor, B.N. (2000) Review of Modern Physics, 72, 351-495.

12.   (Yearly) NASA Extragalactic Database.

Chapter 6

MODIFIED NEWTONIAN DYNAMICS AS AN ENTROPIC FORCE

Diego A. Carranza and Sergio Mendoza
Instituto de Astronoma, Universidad Nacional Autónoma de México, México D.F., México

ABSTRACT

Under natural assumptions on the thermodynamic properties of space and time with the holographic principle, we reproduce a MOND-like behaviour of gravity on particular scales of mass and length, where Newtonian gravity requires a modification or extension if no dark matter component is introduced in the description of gravitational phenomena. The result is directly obtained with the assumption that a fundamental constant of nature with dimensions of acceleration needs to be introduced into gravitational interactions. This in turn allows for modifications or extensions of the equipartion law and/or the holographic principle. In other words, MOND-like phenomenology can be reproduced when appropriate generalised concepts at the thermodynamical level of space and/or at the holographic principle are introduced. Thermodynamical modifications are reflected in extensions to the equipartition law which occur when the temperature of the system drops below a critical value, equals to Unruh's temperature evaluated at the acceleration constant scale introduced for the description of the gravitational phenomena. Our calculations extend the ones by [1] in which Newtonian gravity is shown to be an emergent phenomenon, and together with it reinforces the idea that gravity at all scales is emergent.

INTRODUCTION

The laws for black hole mechanics have suggested a remarkable similarity with the three laws of thermodynamics, in such a way that quantities associated to black hole properties have their corresponding thermodynamic equivalent interpretation [2] -[4] . In particular, the black hole area—which is determined by its horizon—is related to the associated black hole entropy, in the sense that it cannot decrease in time under any physical process on a closed system. The temperature of the black hole is given by the Hawking-Zeldovich temperature and is inversely proportional to the mass of the black hole [5] [6] . The well known interpretation of entropy as a quantity that offers a measure of non-available information—or disorder—in a system, has leaded directly to the idea that the increase in entropy, and therefore in area, is due to the loss of information. This is due to the fact that when a particle crosses the horizon it has no more causal relation with the rest of the universe [2] .

All the above suggest the possibility for a deep relation between thermodynamics and gravity. This has been studied mainly in the relativistic regime under the concept of emergent gravity, considering thermodynamics as a more fundamental theory from which, general relativity can be derived (see e.g. [7] and references therein). Using a metric treatment of the thermodynamic variables in a curved space-time, [8] has been able to derive Einstein's field equations. In the non-relativistic regime, [1] used very simple assumptions about space, energy and information in order to show that the first law of thermodynamics, along with an entropy formula, leads directly to Newton's law of gravity.

In recent years, a growing number of independent observations have suggested that gravity requires modification [9] - [14] and not the inclusion of unknown dark matter entities. All these facts are encouraging, since similar arguments like Verlinde's ones can be used to search for a more profound fundamental basis for an extended theory of gravity like the one proposed by [15] . This approach is based on Buckingham's Π theorem of dimensional analysis, and deals with the problem of a test particle under a gravitational field generated by a central point mass M in an extended regime of gravity. Such analysis gives the general form for the acceleration a experienced by a test particle at a distance r from the central mass:

$$a = a_0 f(x), \quad \text{with } x := \frac{l_M}{r} := \frac{1}{r}\left(\frac{GM}{a_0}\right)^{1/2},$$

(1)

where a_0 is Milgrom's acceleration constant [16] , introduced as an extra fundamental constant of nature when extended gravitational phenomena is described. The unknown dimensionless function $f(x)$ depends only on the dimensionless quantity x, which according to the Π theorem, is the key parameter in a description of extended gravity. The function $f(x)$ is such that: $f(x) \to x^2$ when $x \gg 1$, converging to Newtonian gravity; and $f(x) \to x$ when $x \ll 1$, reaching the deep MONDian regime of gravity. The authors also showed that the transition function could be written as:

$$f(x) = x \frac{1 \pm x^{n+1}}{1 \pm x^n},$$

(2)

where n is a constant that must be fixed via astronomical observations. A large value of n means that the function $f(x)$ abruptly changes from a MONDian to a Newtonian regime of gravity about $x \approx 1$. Small values of n yield soft transitions about the same point.

For the case of spherically symmetric mass distribution, this extended Newtonian gravity approach proposal reproduces a MOND-like phenomenology [16] and has proven to be in good agreement with observations in astrophysical systems across different scales without invoking any dark matter component [9] [10] [15] [17] -[19] .

In another attempt to obtain the MOND-like force formula, [20] have followed Verlinde's work assuming that the equipartition law of energy is modified. In this approach, a Debye's function is introduced, and becomes identified as the MOND interpolation function $\mu(a/a_0)$. The reason of doing so is simply to satisfy Newtonian and MONDian regimes of gravity at their corresponding limits. Furthermore, they also give an expression for the value of Milgrom's acceleration constant in terms of Debye's temperature T_D :

$$a_0 = \frac{12 c k_B T_D}{\pi \hbar},$$

(3)

thus interpreting it as a cut off temperature below which, modifications to the equipartition law must occur. This approach takes into account the fact that the dynamical sector needs to be modified and so, the validity of Newton's law of gravitation remains unaffected. As explained by the dimensional analysis of [15] , the modification must occur in the gravitational sector and rather than working with a transition function $\mu(a/a_0)$, the extension is carried out through the inclusion of a transition function $f(x)$.

In this work, we show how, using arguments about thermodynamics and information, it is possible to derive in several ways an equation for the gravitational force in an extended modified gravity regime, which supports the idea that gravity can be understood as an emergent force, i.e. a consequence of deeper fundamental principles. The article is organised as follows: in Section 2 we review the main hypothesis made by [1] , and then we use dimensional analysis arguments to study modifications to the equipartition law of energy via two approaches, one of them purely thermodynamic, and the other only gravitational. Using another point of view, Section 3 is devoted to study possible modifications to the holographic principle, specifically, to the number of bits contained inside a screen under the assumption that Milgrom's acceleration constant a_0 is a fundamental constant of nature. Finally, in Section 4 we discuss our main results.

MODIFICATIONS TO THE EQUIPARTITION LAW

We begin this section reviewing briefly some of the main ideas and hypothesis made in Verlinde's work [1]. One of the most important assumptions made by him, is that the information describing a physical system is stored on spatial surfaces, or screens, that are ruled by the holographic principle. Every surface behaves as a "stretched horizon" of a black hole (although in this case it has no physical properties like density or surface pressure), and when a particle interacts with it, the entropy, and consequently, the amount of information gets affected. In principle we do not know the shape of the surfaces, so for simplicity we can consider each screen as closed and spherical with radius r. Each surface contains N bits of information. One can also think that on each fundamental Planck square area, the maximum information that can be stored is one bit. This length is constructed with three fundamental constants of nature: 1) Newton's constant of gravity G, 2) the velocity of light c and 3) Planck's constant $\hbar$. With these assumptions, the number of bits N stored on each screen can be expressed as:

$$N = \frac{A}{l_P^2} = \frac{4\pi r^2}{l_P^2},$$
(4)

being A the area of the spherical screen and $l_P^2 = G\hbar/c^3$ the Planck area.

The main motivation by [1] to think of gravity as a force related to entropy has its origin on the restitutive force that acts on a polymer when it suffers a displacement Δx. This force tends to restore the polymer to its original position since this configuration maximises the entropy. The link with gravitation consists on a similar idea, for which there is an entropic force that emerges as a consequence of the system searching for a configuration of maximum entropy when a particle approaches a given screen. We assume that inside the screen the dynamics allow us to define energy, and consequently the associated mass M and temperature T are well defined quantities. With this, we can use the first law of thermodynamics to find the force F associated to changes in the stored information, i.e. due to a change in entropy ΔS given by:

$$F\Delta x = T\Delta S,$$
(5)

for a constant volume. Let us now find the expression for the gravitational force by considering gravity as an entropic force. For this, we follow the approach by [1] [8] analysing the behaviour of a test mass m particle near a black hole horizon. At a distance of one Compton length from the horizon, the particle can be considered to be part of the black hole and so, its entropy is increased in the following way:

$$\Delta S = 2\pi k_{\mathrm{B}} \frac{mc}{\hbar} \Delta x,$$

(6)

when a displacement Δx occurs. In other words, a change in the particle position causes an increment on the system's entropy, such that it tries to maximise it and as such, the horizon can be substituted by a screen. The other key assumption to make is that the energy contained inside the surface satisfies the principle of equipartition, and that it is equal to Mc^2, where M is its associated mass. At this point, we take an approach similar to the one followed by [20] , i.e. we search for a modification of the equipartition law. This can be achieved with the help of Buckingham's theorem of dimensional analysis, since it provides a way to find the general form for the energy when modifying effects are considered.

To do so, note that the dimensional relevant quantities of the problem are the energy E associated to the screen, it's temperature T—or more important for a dimensional analysis treatment its energetic temperature $k_{\mathrm{B}}T$ (where k_{B} is Boltzmann's constant), the speed of light c and Planck's constant $\hbar$. Since we are studying a gravitational problem in an extended regime, Milgrom's constant a_0 is also introduced, along with Newton's constant of gravity G. The associated mass M of the spherical screen with radius r is given by $E = Mc^2$. In other words, seven physical quantities $(k_{\mathrm{B}}T, c, \hbar, G, a_0, M, r)$ play a fundamental role in the description of the energy E of a screen. Since there are three independent dimensions (length, time and mass), Buckingham's Π theorem demands the energy of the screen to be given by [21] :

$$E = \frac{1}{2} N k_{\mathrm{B}} T F\left(x, \frac{T}{T_*}, \frac{\lambda}{r}, \frac{l_{\mathrm{p}}}{r} \right),$$

(7)

where F is an unknown function of four dimensionless parameters, x was defined in Equation (1), $T_* := a_0 \hbar / c k_{\mathrm{B}}$ is a quantity with dimensions of temperature—which coincides to the Unruh temperature evaluated at the acceleration a_0—and $\lambda := c^2 / a_0$ is a characteristic length that appears when relativistic effects are considered in the extended regime of gravity [17] . The dimensionless factor $N/2$ has been introduced for consistency with the standard law of equipartition. It is important to note that the temperature T_* is proportional to the so called "cut off Debye temperature" given in Equation (3) and constitutes a characteristic temperature scale in an extended regime of gravity where MONDian effects are to be taken in consideration.

The ratio l_P/r appears as a characteristic dimensionless quantity in Equation (7), but Equation (4) suggests that this quantity is closely related to the holographic principle and not to a modification to the equipartition law. Also, since that ratio does not contain a_0 it will be incapable to account for any MOND phenomenology. In other words, the equipartition energy E is only a function of three dimensionless parameters: x, T/T_* and λ/r. The simplest assumption to make for the function F is that it is of power-law form on any of its arguments, i.e.:

$$E = \frac{\xi}{2} N k_B T x^\alpha \left(\frac{T}{T_*}\right)^\beta \left(\frac{\lambda}{r}\right)^\gamma,$$

(8)

where ξ is a constant of proportionality.

In the remaining of this section, we study three separate cases associated with the previous relation:

Case (A)

Let us consider $\alpha = \gamma = 0$ in Equation (8) to obtain:

$$E = \frac{\xi}{2} N k_B T \left(\frac{T}{T_*}\right)^\beta.$$

(9)

The physics behind this choice of parameters can be understood under the basis of the case studied by [20] since, as it was mentioned previously, the temperature T_* corresponds—except for a 2π factor—with the Unruh temperature on a holographic screen. In that particular study, the authors dealt with a non-standard equipartition law of energy including the one dimensional Debye function to take into account the corrections at low temperatures. Their equipartition law corresponds to (9), with T representing the one dimensional Debye function.

Equating relation (9) to Mc^2, and using Equation (4), the temperature can be written as:

$$T = \frac{\hbar}{ck_B} \left(\frac{a_0^\beta GM}{2\pi \xi r^2}\right)^{1/(\beta+1)}.$$

(10)

Substituting this into (5), and employing (6), as made by Verlinde, the resultant entropic force is:

$$F = 2\pi m \left(\frac{a_0^\beta GM}{2\pi \xi_1 r^2} \right)^{1/(\beta+1)}.$$

(11)

Taking $\beta = 0$ and $\xi = 1$ in Equation (9), the standard equipartition law is obtained and as seen from relation (11), Newton's gravitational law is recovered. In order to obtain a MONDian $1/r$ gravitational force law we must take $\beta = 1$ and $\xi = 2\pi$:

$$F = m \frac{\sqrt{a_0 GM}}{r}.$$

(12)

In other words, the equipartition energy must satisfy the following condition:

$$E = \frac{1}{2} N k_B T \begin{cases} 1, & \text{for Newtonian gravity,} \\ 2\pi T/T_* , & \text{for MOND-like gravity.} \end{cases}$$

(13)

Case (B)

Let us now consider the case when $\beta = \gamma = 0$ in Equation (8).

Once again, using the equivalence between mass and energy, and the explicit form of Planck's area, we can write:

$$Mc^2 = \xi \frac{2\pi r^2 c^3}{G\hbar} k_B T x^\alpha,$$

(14)

which with the aid of Equations (5) and (6), gives following expression for the entropic force:

$$F = \frac{GmM}{\xi x^\alpha r^2}.$$

(15)

The choice $\alpha = 0$ and $\xi = 1$ results into Newton's law of gravity and $\alpha = 1$ with $\xi = 1$ converge to MOND's force Formula (12).

The acceleration exerted on the test mass m is then given by:

$$a = \frac{a_0}{r^2} \frac{GM}{a_0} \frac{1}{x^\alpha} = a_0 \frac{x^2}{x^\alpha}.$$

(16)

It is convenient to express it in this form, since we can easily compare it with (1). It can be observed that given the form proposed for the energy E, in the Newtonian regime, the ratio $x^2/x^\alpha \rightarrow x^2$, and in the modified one, $x^2/x^\alpha \rightarrow x$, which are the same limits satisfied by the function $f(x)$, as discussed in Section 1. Since the transition function $f(x)$ is given by Equation (2) it follows that the equipartition energy can be written in a very general form as:

$$E = \frac{1}{2} N k_{\mathrm{B}} T x \frac{1 \pm x^n}{1 \pm x^{n+1}}$$

(17)

Case (C)

Finally, we study the case for which $\alpha = \beta = 0$ in Equation (8) is considered. Equating the resulting equation to Mc^2 it is possible to find the temperature in terms of the mass and distance and so with the aid of relation (5) it follows that:

$$F = G \frac{mM}{\xi r^2} \left(\frac{a_0 r}{c^2} \right)^\gamma .$$

(18)

The choice $\gamma = 0$ and $\xi = 1$ yields Newton's gravitational law. A MOND-like limit $1/r$ seems to be possible when $\gamma = 1$, but the square root dependence on the mass cannot be achieved with this choice. Furthermore, the speed of light still appears on this non-relativistic limit. This all means that the dimensionless parameter λ/r must not appear on Equation (8), which means that $\gamma = 0$.

To summarise this section, note that the inclusion of Milgrom's acceleration constant as a fundamental quantity of nature related to gravitational phenomena, is capable of generalising the equipartition law in such a way that either Newton or MOND force formulae can be obtained. This result reinforces the idea that, at all scales, gravity is an emergent force with a thermodynamic nature.

MODIFICATIONS TO THE HOLOGRAPHIC PRINCIPLE

Let us now search for a MOND-like force formula assuming that the equipartition law has its usual form and allowing for modifications related to the way in which information is stored on the holographic screens (cf. Equation (4)). The only additional constant that needs to be taken into consideration is a_0 and so, Buckingham's theorem of dimensional analysis requires that the number of bits N stored on a screen is given by:

$$N = \frac{4\pi r^2}{l_{\mathrm{P}}^2} g(x),$$

(19)

where $g(x)$ is an unknown dimensionless function. Since Equation (19) is a generalisation of the standard relation (4), the function $g(x) \to 1$ in the

Newtonian regime of gravity, i.e. for $x \gg 1$. The previous equation assumes the validity of the holographic principle and only changes the quantitative way in which information is stored on a particular screen. Let us assume that the function $g(x)$ is a power law, i.e. $g(x) = x^{\varsigma}$, to obtain:

$$N = \frac{4\pi r^2}{l_{\mathrm{p}}^2} x^{\varsigma},$$

(20)

With this in mind, we can follow Verlinde's analysis. Assuming the validity of the equipartition law of energy, and the equivalence between mass and energy inside the screen, along with Equation (20), it follows that:

$$T = \frac{GM\hbar}{2\pi c k_B r^2 x^{\varsigma}}.$$

(21)

Direct substitution of the previous equation on relation (6), with the aid of the first law of thermodynamics (5) yields the following expression for the acceleration caused by this entropic force:

$$a = \frac{GM}{r^2 x^{\varsigma}}.$$

(22)

The choice $\varsigma = 1$ yields a MONDian gravitational force and $\varsigma = 0$ a Newtonian one. Comparison of the previous equation with (1) it follows that the complete transition (19) is given by:

$$N = \left(\frac{4\pi r^2}{l_{\mathrm{p}}^2}\right) x \frac{1 \pm x^n}{1 \pm x^{n+1}}.$$

(23)

Finally, a possible alternative way to modify Verlinde's result it to combine these two approaches, i.e. assume that both the equipartition law of energy and the holographic principle get modified when MONDian effects are introduced. Based on the previous analysis, this can be studied if we introduce the parameter x as follows:

$$E = \frac{1}{2} N k_B T p(x), \quad N = \frac{4\pi r^2}{l_{\mathrm{p}}^2} q(x),$$

(24)

where $p(x)$ and $q(x)$ are unknown functions of x. Given that both p and q are proportional to E and N respectively, they will be inversely proportional to the gravitational acceleration, i.e.:

$$a = \frac{GM}{r^2} \frac{1}{p(x)q(x)} = a_0 \frac{x^2}{p(x)q(x)},$$

(25)

and so,

$$p(x)q(x) = \frac{x^2}{f(x)},$$

(26)

according to Equation (1) with $p(x)$ and $q(x)$ tending to 1 in the Newtonian limit, i.e. when $x \gg 1$. Also, the previous equation imposes the following restriction: $p(x)g(x) \to x$ in the MONDian regime of gravity, i.e. $x \ll 1$.

DISCUSSION

As explained by [22] [23] and in a more profound and empirical way by [24] , if gravitational phenomena require to be modified at a certain scales of mass and length one needs to incorporate a new fundamental constant of nature relevant to all gravitational phenomena at those scales. This gravitational constant is as important as Newton's constant of gravity and can be mathematically manipulated as to have dimensions of acceleration which converge to Milgrom's acceleration constant a_0. This is so since gravitational phenomena do not follow the standard Newtonian (or general relativistic) behaviour of gravity at scales which greatly differ from the ones in which precise gravitational experiments have been performed to test the validity of Newton's law of gravitation (or Einstein's general relativity—cf. [25]). The behaviour of gravity at those scales can be considered as independent of the behaviour of standard gravity and as such a new fundamental constant of nature has to be introduced into the description of gravitational phenomena, which is a standard procedure to follow when extensions of a particular physical theory are performed (e.g. [21]).

In this article we have introduced this extra fundamental constant of nature a_0 in the description of gravity and used thermodynamic and information properties of space and time in order to show that a MONDian force law can be obtained by assuming the validity of the holographic principle. Specifically, this has been studied under two different approaches using Buckingham's Π theorem of dimensional analysis:

1)   In the first approach, the equipartition law is modified and the holographic principle keeps its standard form, resulting on a temperature scale T_* (corresponding to the Unruh temperature evaluated at the Milgrom acceleration a_0). This temperature corresponds to the one already found by [20] who studied the problem of the emergence of gravity modifying the equipartition law using a Debye model. [26] also worked on modifications focused on thermodynamical statistical properties and in the same way as [20] dynamical modifications with its corresponding

MOND transition function $\mu(a/a_0)$ were obtained. As explained by [15], direct dimensional analysis points towards a modification of the gravitational force, and the calculations we performed in this work were done by keeping Newton's second law untouched, allowing for an extension on the gravitational sector.

2)   In a second approach, modifications to the holographic principle—with the equipartition law unchanged— have showed to be able to explain in a natural way how gravity transits from a Newtonian regime to a modified one.

The important point about these two different ways is that they are consistent with the general formula for acceleration experienced by a particle under a gravitational field given in (1). In this sense, and in the context of emergent gravity, it suggests that the transition observed across different astrophysical systems could be a consequence of a modification at a deeper level in the equipartition law and/or in the holographic principle, i.e. the observed effects at large scales in gravitational systems reflect the behaviour of physical laws at a deeper thermodynamical level. More generally, it has been also considered the possibility of modifications of both the holographic principle and the equipartition law in such regimes.

As pointed by [9], observations of globular clusters yield a lower limit on the exponent $n \gtrsim 8$ in Equation (2), meaning that the transition function $f(x)$ is quite abrupt, i.e. $f(x) = x$ for $x \leq 1$ (MONDian regime) and $f(x) = x^2$ for $x \geq 1$ (Newtonian regime), with almost no soft transition from one regime to the other. This means that the transition functions E, N, $p(x)$ and $q(x)$ calculated in this article must present a rather abrupt transition.

A full non-relativistic theory of gravity can be constructed assuming a modification of inertia as described by [22], but as shown in this work the modification naturally appears in the force sector and not on the dynamical one.

With a few natural assumptions about space and information, the main result of this article is to show that gravity can be considered an emergent phenomenon also in the MONDian regime. This suggests that the force of gravity on this extended regime is not a fundamental force of nature, but a consequence of the inherent properties of space and time. Since [1] showed that Newtonian gravity emerges from the thermodynamic properties of space and time, this all suggests that gravitation is an emergent phenomenon at all scales of mass and length.

ACKNOWLEDGEMENTS

We thank an anonymous referee for his fruitful comments on the first version of this article. This work was supported by a DGAPA-UNAM grant (PAPIIT IN111513-3) and a CONACyT grant (240512). DAC and SM thank support granted by CONACyT 480147 and 26344. The authors gratefully acknowledge the comments made by Ehoud Pazy and Hristu Culetu for the valuable comments made of an earlier version of this article.

REFERENCES

1. Verlinde, E. (2011) Journal of High Energy Physics, 4, 29.http://dx.doi.org/10.1007/JHEP04(2011)029

2. Bekenstein, J.D. (1973) Physical Review D, 7, 2333-2346.http://dx.doi.org/10.1103/PhysRevD.7.2333

3. Bekenstein, J.D. (1974) Physical Review D, 9, 3292-3300.http://dx.doi.org/10.1103/PhysRevD.9.3292

4. Bardeen, J.M., Carter, B. and Hawking, S.W. (1973) Communications in Mathematical Physics, 31, 161-170. http://dx.doi.org/10.1007/BF01645742

5. Hawking, S.W. (1974) Nature (London), 248, 30-31. http://dx.doi.org/10.1038/248030a0

6. Townsend, P.K. (1997) Black Holes. ArXiv General Relativity and Quantum Cosmology e-Prints.

7. Padmanabhan, T. (2010) Reports on Progress in Physics, 73, Article ID: 046901.http://dx.doi.org/10.1088/0034-4885/73/4/046901

8. Jacobson, T. (1995) Physical Review Letters, 75, 1260-1263.http://dx.doi.org/10.1103/PhysRevLett.75.1260

9. Hernandez, X. and Jiménez, M.A. (2012) Astrophysical Journal, 750, 9.http://dx.doi.org/10.1088/0004-637X/750/1/9

10. Hernandez, X., Jiménez, M.A. and Allen, C. (2012) European Physical Journal C, 72, 1884.http://dx.doi.org/10.1140/epjc/s10052-012-1884-6

11. Mastropietro, C. and Burkert, A. (2008) Monthly Notices of the Royal Astronomical Society, 389, 967-988. http://dx.doi.org/10.1111/j.1365-2966.2008.13626.x

12. Lee, J. and Komatsu, E. (2010) The Astrophysical Journal, 718, 60-65. http://dx.doi.org/10.1088/0004-637X/718/1/60

13. Thompson, R. and Nagamine, K. (2012) Monthly Notices of the Royal Astronomical Society, 419, 3560-3570. http://dx.doi.org/10.1111/j.1365-2966.2011.20000.x

14. Moffat, J.W. and Toth, V.T. (2010) Can Modified Gravity (MOG) Explain the Speeding Bullet (Cluster)? http://arxiv.org/abs/1005.2685

15. Mendoza, S., Hernandez, X., Hidalgo, J.C. and Bernal, T. (2011) Monthly Notices of the Royal Astronomical Society, 411, 226-234. http://dx.doi.org/10.1111/j.1365-2966.2010.17685.x

16. Milgrom, M. (1982) The Astrophysical Journal, 270, 371-389. http://dx.doi.org/10.1086/161131

17. Bernal, T., Capozziello, S., Hidalgo, J.C. and Mendoza, S. (2011) European Physical Journal C, 71, 1794. http://dx.doi.org/10.1140/epjc/s10052-011-1794-z

18. Carranza, D.A., Mendoza, S. and Torres, L.A. (2013) European Physical Journal C, 73, 2282. http://dx.doi.org/10.1140/epjc/s10052-013-2282-4

19. Mendoza, S., Bernal, T., Hernandez, X., Hidalgo, J.C. and Torres, L.A. (2013) Monthly Notices of the Royal Astronomical Society, 433, 1802-1812. http://dx.doi.org/10.1093/mnras/stt752

20. Sheykhi, A. and Sarab, K.R. (2012) Journal of Cosmology and Astroparticle Physics, 2012, 12. http://dx.doi.org/10.1088/1475-7516/2012/10/012

21. Sedov, L.I. (1959) Similarity and Dimensional Methods in Mechanics. Academic Press, Waltham.

22. Famaey, B. and McGaugh, S.S. (2012) Living Reviews in Relativity, 15, 10. http://dx.doi.org/10.12942/lrr-2012-10

23. Mendoza, S. (2011) Extending Cosmology: The Metric Approach. http://arxiv.org/abs/1208.3408

24. Mendoza, S. and Olmo, G. (2012) Living Reviews in Relativity, 15, 10.

25. Will, C.M. (1992) Theory and Experiment in Gravitational Physics. Cambridge University Press, Cambridge.

26. Pazy, E. and Argaman, N. (2012) Physical Review D, 85, Article ID: 104021. http://dx.doi.org/10.1103/PhysRevD.85.104021

Chapter 7

THE GENERALIZED NEWTON'S LAW OF GRAVITATION VERSUS THE GENERAL THEORY OF RELATIVITY

Arbab Ibrahim Arbab
Department of Physics, Faculty of Science, University of Khartoum, Khartoum, Sudan

ABSTRACT

Einstein general theory of relativity (GTR) accounted well for the precession of the perihelion of planets and binary pulsars. While the ordinary Newton law of gravitation failed, a generalized version yields similar results. We have shown here that these effects can be accounted for as due to the existence of gravitomagnetism only, and not necessarily due to the curvature of space time. Or alternatively, gravitomagnetism is equivalent to a curved space-time. The precession of the perihelion of planets and binary pulsars may be interpreted as due to the spin of the orbiting planet (m) about the Sun (M). The spin (S) of planets is found to be related to their orbital angular momentum (L) by a simple formula, viz., $S \propto \frac{m}{M} L$.

INTRODUCTION

We have recently introduced gravitomagnetism as a true cause of the precession of the perihelion of the orbit of planets and binary pulsars [1]. Einstein attributed these effects to the curvature of space-time. The effect of gravitomagnetism, in a similar manner to electromagnetism, is the Larmor precession of a gravitational moment in the gravitomagnetic field induced by the Sun on the planets.

Le Verrier discovered that the orbital precession of the planet Mercury was not quite what it should be; the ellipse of its orbit precesses by some minute value than the predicted by the Newtonian theory of gravitation, even after all the effects of the other planets had been accounted for [1]. This value amounts to 43 arcseconds per century. Several classical explanations were put forward, e.g., an interplanetary dust, an unobserved oblateness of the Sun, an undetected moon of Mercury, or a new planet named Vulcan. Others suggested that the Newton inverse-square law is not correct, and accordingly proposed a power law with an exponent that slightly differs from 2. Moreover, some authors argued in favor of a velocitydependent potential (see [1] and references there in).

To resolve the above mentioned dilemmas, Einstein used a pseudo-Riemannian geometry to allow for the curvature of space-time that was necessary for the reconciliation of the observed gravitational phenomena. He concluded that the space-time should be curved in order to reproduce the observed physical laws of gravitation. Owing to Einstein's theory of general relativity, particles of negligible mass travel along geodesics in the spacetime. An exact solution to the Einstein field equations is the Schwarzschild metric, which corresponds to the external gravitational field of a stationary, uncharged, nonrotating, spherically symmetric body of mass M. It is characterized by a length scale r_s, known as the Schwarzschild radius. The immediate solutions of the field equations explained the anomalous precession of Mercury, and predicted the observed bending of light, which were later confirm experimentally [2].

On the other hand, the theory of electromagnetic interaction is accomplished by Maxwell. This is coined in the four Maxwell equations relating the electric and magnetic fields to the electric charges and current. Lorentz then obtained the force experienced by a charged particle in electric and magnetic fields. Larmor has found that when an electron (magnetic moment) is placed in an external magnetic field, the magnetic moment precess about the magnetic field direction. This precession is due to the spin of the electron. This effect is prominent in the spinorbit interaction exhibited by hydrogen atom [3,4].

If we now consider gravitation with some scrutiny, we will find that, unlike electromagnetism, moving mass doesn't create a magnetic-like field. Thus, Newton law of gravitation is not like Lorentz law of electromagnetism.

In this sense, gravity and electromagnetism are not analogous and can't be utterly compared with gravitation. To remedy this problem, we have to look for a gravitomagnetism counterpart of gravity. In this way, we can say gravity is analogous to electricity and gravitomagnetism is analogous to magnetism. The question is what is the gravitomagnetic field? By analogy, this should be obtained by looking at Biot-Savart law that defines the magnetic field of a uniformly moving charged particle in an electric field. To complete the analogy the charge of the particle should correspond to the mass of the particle. In this way, we may call the electric charge, the electric mass in contrast with the gravitational mass. This furnishes the complete analogy between gravitation and electromagnetism.

How we then avail the electrical phenomena and rules in one paradigm to interpret the other? To answer this question, we have to trust (beforehand) the existing analogies, and base all our new interpretation of the gravitational phenomena by explaining their corresponding ones. In this manner, the precession of the perihelion of the orbit of planets and binary pulsars is obtained from the precession of the electron (magnetic moment) in an external magnetic field. Planets and binary pulsars precess when they experience a gravitomagnetic filed (if any). In this case, we use the same laws holding for the counter (analogous) phenomenon, however.

Moreover, the deflection of light by the Sun is explained by using the laws governing the deflection of a charged particle (α-particle) by the nucleus [5]. If we continue in this manner, we may persuade our selves that, to every electromagnetic phenomenon there are gravitomagnetic counter-phenomena. Hence, electromagnetism and gravitomagnetism are same but different aspects of a unified origin.

In this respect, we will find our-selves distracted to interpret the gravitational phenomena as due to the curvature of space-time. We are then not abide by the GTR to interpret our physical world. Or alternatively, we treat the curvature and gravitomagnetism as a same object, or yield the same effects. This can be trusted if we are able to show that the term responsible for the curvature of space-time in Einstein field equations is the same as the that resulting from the influence of gravitomagnetism.

In this paper, we will show that the gravitomagnetism terms in the generalized Newton law of gravitation is the same as the one in the Einstein general field equations. In this way, we upgrade Newton law of gravitation to the general theory of gravitation, but with different predictions. Thus, the correct Newton law of gravitation still works finely, and expresses gravitational phenomena in accordance with observations. Hence, gravity and electromagnetism are governed by unified laws. In Section 2 we present the potential that gives rise to the precession of perihelion in the GTR. We compare this potential with that arising from the gravitomagnetic field.

We find that the gravitomagnetic term is $\frac{\pi}{3}$ of the Einstein term (GTR). Einstein attributed this term to curvature of space.

Can we say that the gravitomagnetism is the cause of Einstein curvature?

Do we still adopt GTR that requires advanced mathematics, as the theory of gravitation and leave the simplyunderstood Newton's laws of gravitation aside? In effect, the gravitomagnetic theory (or Gravitational Lorentz force) is simple and can easily be handled without recourse to tensor (advanced mathematics) analysis to unravel gravitational phenomena. Besides, it is analogous to electromagnetic theory that is well understood and complies utterly with experimental facts. The idea of curvature of space is no longer adopted. Moreover, the Einstein's dream of unification of fundamental forces in nature will become imminent within this framework.

THE GENERAL THEORY OF RELATIVITY (GTR)

Einstein attributed the gravitational phenomena, now known, to the effect of the curvature of space-time induced by the presence of a massive object [2]. The effective gravitational potential of the object of mass m moving around a massive object of mass M takes the form [6]

$$U(r) = -\frac{GMm}{r} + \frac{L^2}{2mr^2} - \frac{GML^2}{c^2mr^3},$$

(1)

and the force, $F = -\frac{\partial U}{\partial r}$, can be written as

$$F(r) = -\frac{GMm}{r^2} + \frac{L^2}{mr^3} - \frac{3GML^2}{mc^2r^4},$$

(2)

where L is the orbital angular momentum of the mass m.

This inverse-cubic energy term in Equation (1) causes elliptical orbits to precess gradually by an angle $\delta\varphi$ per revolution [2]

$$\delta\varphi = \frac{6\pi GM}{c^2 a(1-e^2)},$$

(3)

where e and a are the eccentricity and semi-major axis of the elliptical orbit, respectively. This is known as the anomalous precession of the planet Mercury.

Another prediction famously used as evidence for GTR, is the bending of light in a gravitational field. The deflection angle is given by [2]

$$\delta\theta = \frac{4GM}{c^2 b},$$

(4)

where b is the distance of closest approach of light ray to the massive object.

Therefore, the gravitomagnetic force is equal to $\frac{\pi}{3}$ of the GTR force. Whether, the gravitational phenomena are in full agreement with our gravitomagnetic model or with GTR is a subject of the present and future observations. At any rate, we are lucky to have two complementary paradigms explaining the same effect in different ways. Can we deduce that it is the gravitomagnetic field that curves the space and not the Sun mass? Or can we say that it is the curvature that produces the gravitomagnetism?

THE GENERALIZED NEWTON LAW OF GRAVITATION

We have shown recently that Newton law of gravitation can be written, as a Lorentz-like law, as [7]

$$F(r) = mE_g + mv \times B_g, \quad E_g = a = \frac{v^2}{r},$$

(5)

where

$$B_g = \frac{v \times E_g}{c^2}.$$

(6)

Thomas introduced a factor $\frac{1}{2}$ to account for the spin-orbit interaction in hydrogen atom [8]. Here B_g is measured in s^{-1}. To convert it to rad/sec, we multiply it by 2π. Hence, the gravitomagnetic force becomes

$$F_m(r) = -\frac{\pi m v^4}{c^2 r}, \quad a = \frac{v^2}{r}, \quad v^2 = \frac{GM}{r}.$$

(7)

The gravitomagnetic field is divergenceless, since

$$\nabla \cdot B_g = \frac{1}{c^2} \nabla \cdot (v \times E_g)$$

$$\nabla \cdot B_g = \frac{1}{c^2} E_g \cdot (\nabla \times v) - \frac{1}{c^2} v \cdot (\nabla \times E_g)$$

$$\nabla \cdot B_g = -\frac{1}{c^2} v \cdot \frac{\partial B_g}{\partial t} = -\frac{1}{c^2} \frac{\partial}{\partial t} (v \cdot B_g) = 0.$$

This implies that the gravitomagnetic lines curl around the moving mass (gravitational current) creating it. This may also rule out the existence of negative mass. Therefore, as no magnetic monopole exits; no gravitomagnetic monopole (antigravity) exits. Thus, the search for magnetic monopole is tantamount to that of antigravity.

The angular momentum is defined by $L = mvr$, so that Equation (7) becomes

$$F_m(r) = -\frac{\pi GML^2}{mc^2 r^4}.$$

(8)

The second term in Equation (2) is due to the centrifugal term arising from a central force field. In polar coordinates the force is written as

$$ma = m\left(\ddot{r} - r\dot{\theta}^2\right)\hat{e}_r + m\left(r\ddot{\theta} + 2\dot{r}\dot{\theta}\right)\hat{e}_\theta.$$

(9)

For a central force the second term vanishes. It yields,

$\dot{\theta} = \frac{L}{mr^2}$, so that the first term becomes

$$ma_r = m\ddot{r} - \frac{L^2}{mr^3}.$$

(10)

Substituting Equation (10) in Equation (5) yields the full effective central force, owing to gravitomagnetism, as

$$F(r) = -\frac{GMm}{r^2} + \frac{L^2}{mr^3} - \frac{\pi GML^2}{mc^2 r^4}.$$

(11)

The corresponding potential will be

$$U(r) = -\frac{GMm}{r} + \frac{L^2}{2mr^2} - \frac{\pi GML^2}{3mc^2 r^3}.$$

Comparison of Equations (2) and (11) reveals that the gravitomagnetic force is equal to $\frac{\pi}{3}$ of the curvature force. Consequently, the generalized Newton law of gravitation and the general theory of relativity produce the same gravitational phenomena.

The gravitomagnetic force term, the last term in Equation (11), can be written as

$$\frac{\pi GML^2}{mc^2 r^4} = \frac{\pi G^2 M^2 m}{c^2 r^3}, \quad \text{where,} \quad v^2 = \frac{GM}{r}.$$

(12)

Finally, Equation (11) can be written as

$$F(r) = -\frac{GMm}{r^2} + \frac{J_{eff}^2}{mr^3},$$

(13)

where

$$J_{eff.}^2 = L^2 - \left(\frac{\sqrt{\pi}GMm}{c} \right)^2.$$

(14)

PRECESSION OF PLANETS AND BINARY PULSARS

Owing to the above equivalence between gravitomagnetism and GTR, we interpret the precession of the perihelion of planets and binary pulsars as a Larmor-like precession, and not due to the GTR interpretation as due to the curvature of space-time. We may attribute this precession as due to the precession of gravitational moment (mass) in a gravitomagnetic field induced by the massive objects (Sun). In electromagnetism, the Larmor precession is defined by [4]

$$\omega = \frac{e}{2m} B,$$

(15)

while in gravitation (since B_g is in s^{-1} and $e \Leftrightarrow m$) it is defined as [1]

$$\omega_g = 2\pi \left(\frac{B_g}{2} \right) = \frac{\pi v^3}{rc^2}, \quad B_g = \frac{va}{c^2} = \frac{v^3}{rc^2},$$

(16)

where (ω_g is in rad/seec) and

$$a = \frac{v^2}{r}$$

(17)

The precession rate in Equation (16) can be written as

$$\omega_g = \pi \left(\frac{2\pi GM}{Tc^2 r} \right) = \frac{\delta\phi_g}{T},$$

(18)

where $T = \frac{2\pi r}{v}$ is the period of revolution. This corresponds to a precession angle of

$$\delta\varphi_g = \pi \left(\frac{2\pi GM}{c^2 r} \right) \text{ rad/s},$$

(19)

that is equal to $\frac{\pi}{3}$ of the curvature effect, and for elliptical orbit $r = a(1 - e^2)$.

DEFLECTION OF α -PARTICLES BY THE NUCLEUS

We would like here to interpret the deflection of light by the Sun gravity in an analogous way to the deflection of α -particles by the nucleus, without resorting to the GTR calculation. The deflection angle of α-particles by a nucleus is given by [5]

$$\Delta\theta_e = \frac{4keQ}{mbv^2}$$

(20)

where Q is the nucleus charge, v the α-particle speed, k Coulomb constant, and b the impact factor. The corresponding gravitational analog for the deflection of light will be, $v \to c$, $e \to m$, $Q \to M$, $k \to G$, [9]

$$\Delta\theta_g = \frac{4GM}{bc^2}$$

(21)

without resorting to GTR calculation. Recall that, according to Equivalence Principle, all particles in gravity accelerate without reference to their mass (whether massive or massless). Therefore, it doesn't matter whether light has a mass or not. The relation in Equation (21) is the same as the relation obtained by GTR as in Equation (4). The minimum distance α particles can approach the nucleus is given by equating the kinetic energy and the Coulomb potential energy that yields the relation

$$b_e = \frac{2kq_1q_2}{mv^2}.$$

(22)

In gravitation and for light scattered by the Sun gravity, the above relation gives ($q_1 \to m$, $q_2 \to M$ and $k \to G$)

$$b_g = \frac{2GM}{c^2}.$$

(23)

This is nothing but the Schwarzschild distance that no particle can exceed. Therefore, the complete analogy between gravitation and electricity is thus realized. In this context, we have shown recently that the Larmor dipole radiation has a gravitational analogue [10]. Similarly, the same analogy exists between hydrodynamics and electromagnetism [11].

THE SPIN OF PLANETS

The discovery of the spin of the electron by Goudsmit and Uhlenbeck in 1926 was crucial in understanding many physical phenomena that wouldn't have been explained without [12]. This spin is theoretically formulated by Dirac confirming the experimental finding. However, the spin of planets had been known since long time (1851) that was demonstrated by Foucault's pendulum. In a recent paper we have introduced the gravitomagnetism produced by moving planets as the magnetic field produced by moving charge [1]. We then obtained the gravitational Ampere's and Faraday's laws of gravitomagnetism. The gravitomagnetic moment of a planet due to its orbital motion is given by [1]

$$\mu_L = \frac{v^3 r^2}{2G}.$$

(24)

For circular orbit, Equation (24) yields

$$\mu_L = \left(\frac{M}{2m}\right)L.$$

(25)

In a similar manner the gravitomagnetic moment due to spin will be twice the above value (analogous to electromagnetism)

$$\mu_S = g_S\left(\frac{M}{2m}\right)S,$$

(26)

where g_S defines some gyro-gravitomagnetic ratio that is independent of the planet's mass. If we assume the precession of planets is a spin-orbit interaction, then we can equate $-\mu_S B_g$ (assuming the angle to be zero) to the potential term arising from the gravitomagnetic force in Equation (11). This yields, for circular orbit,

$$S = \left(\frac{4\pi}{3g_S}\frac{m}{M}\right)L, \quad S = \left(\frac{4\pi}{3g_S}\frac{Gm^2}{v}\right).$$

(27)

This is a very interesting equation, since it determines the spin of planets from their orbital angular momentum. With the help of the above equation, the moment of inertia of planets can be precisely determined. It then follows that the spin and the geometrical form of planets is a consequence of its dynamics. Consequently, the spin angular momentum is no longer an intrinsic property of the planet. The energy corresponding to this interaction may be converted into internal energy (heat) inside the planet.

Owing to Equation (27) we are entitled to say that any orbiting planet must spin! Thus, any gravitating object in curvilinear motion must spin. For consistency of the spin of the Earth with the present value with take $g_S = 57$. From this law the moment of inertia of all gravitation objects can be precisely determined. Table 1 shows the anticipated values for the spin and the corresponding moment of inertia of the planetary system. Equation (27) can be used to estimate the hidden central mass around which another mass orbits. It can be generally useful in many astrophysical applications.

Table 1: The predicated values for spin and moment of inertia owing to Equation (27) with $g_s = 57$. Any deviation from known values that may appear could be attributed to the uncertainty in determining the radii of planets. Alternatively, the angle between L and S will be of importance

Planet	Spin (J_z)	Moment of inertia (kg·m²)
Mercury	1.12E+31	8.98E+36
Venus	3.31E+33	1.10E+40
Earth	5.84E+33	8.02E+37
Mars	8.32E+31	1.17E+36
Jupiter	1.35E+39	7.69E+42
Saturn	1.64E+38	1.00E+42
Uranus	5.44E+36	5.43E+40
Neptune	9.45E+36	8.12E+40

CONCLUSION

We have shown that the gravitomagnetism and the general theory of relativity are two theories of the same phenomenon. This entitles us to fully accept the analogy existing between electromagnetism and gravity. Hence, electromagnetism and gravity are unified phenomena. The precession of the perihelion of planets and binary pulsars may be interpreted as a spin-orbit interaction of gravitating objects. The spin of a planet is directly proportional to its orbital angular momentum and mass weighted by the Sun's mass. Alternatively, the spin is directly proportional to the square of the orbiting planet's mass and inversely proportional to its velocity.

ACKNOWLEDGEMENTS

I would like to thank Prof. G. Gadzirayi Nyambuyathe for critically revising the manuscript and drawing my attention to consider a formula for the spin of planets.

REFERENCES

1.   A. I. Arbab, "The Gravitomagnetism: A Novel Explanation of the Precession of Planets and Binary Pulsars," Astrophysics and Space

Science, Vol. 330, No. 1, 2010, pp. 61-68.doi:10.1007/s10509-010-0353-7

2. S. Weinberg, "Gravitation and Cosmology," John Wiley, New York, 1971.

3. D. Griffiths, "Introduction to Electrodynamics," PrenticeHall, Upper Saddle River, 1999.

4. H. L. Malcolm, "Spin Dynamics," John Wiley, New York, 2000.

5. E. Rutherford, "The Scattering of α and β Particles by Matter and the Structure of the Atom," Philosophical Magazine, Vol. 92, No. 4, 2012, pp. 379-398.

6. T. Cheng, "Relativity, Gravitation, and Cosmology," Oxford University Press, Oxford, 2005, p. 108.

7. A. I. Arbab, "The Generalized Newton's Law of Gravitation," Astrophysics and Space Science, Vol. 325, No. 1, 2010, pp. 37-40. doi:10.1007/s10509-009-0145-0

8. L. H. Thomas, "The Motion of the Spinning Electron," Nature, Vol. 117, No. 2945, 1926, p. 514. doi:10.1038/117514a0

9. A. I. Arbab, "A Phenomenological Model for the Precession of Planets and Bending of Light," Astrophysics and Space Science, Vol. 325, No. 1, 2010, pp. 41-45.doi:10.1007/s10509-009-0146-z

10. A. I. Arbab, "On the Gravitational Radiation of Gravitating Objects," Astrophysics and Space Science, Vol. 323, No. 2, 2010, pp. 181-184. doi:10.1007/s10509-009-0058-y

11. A. I. Arbab, "The Analogy between Electromagnetism and Hydrodynamics," Physics Essays, Vol. 24, No. 2, 2011, pp. 254-259. doi:10.4006/1.3570825

12. S. Goudsmit and G. E. Uhlenbeck, "Over Het Roteerende Electron en de Structuur der Spectra," Physica, Vol. 6, 1926, pp. 273-290.

Chapter 8

MIMICKING GENERAL RELATIVITY WITH NEWTONIAN DYNAMICS

E. Goulart

Instituto de Cosmologia Relatividade Astrofisica (ICRA-Brasil/CBPF) Rua Dr. Xavier Sigaud, 150, 22290-180 Rio de Janeiro, RJ, Brazil

ABSTRACT

The aim of this paper is two folded. (1) Showing that Newtonian mechanics of point particles in static potentials admits an alternative description in terms of effective riemannian spacetimes. (2) Using the above geometrization scheme to investigate aspects of the gravitational field as it appears in the Einstein theory. It is shown that the mechanical $(3+1)$ effective metrics are quite similar to Gordon's metric, as it is suggested by the well-known optical-mechanical analogy. Some special potentials are worked out.

INTRODUCTION

The last years have witnessed a great interest in the study of analog gravity [1–14]. The underlying idea is to investigate kinematical aspects of Einstein's theory using physical situations where a kind of geometrization procedure can also be applied. The main motivation to the relativist is the fact that it is possible to assign an effective spacetime to describe propagation features of different physical systems. The generality behind these effective metric techniques is being investigated in diverse contexts such as hydrodynamics, optics inside media, nonlinear field theory, superfluids, and quantum condensates (see [15, 16] for a complete review).

Although the first analog models appeared in the early days of general relativity [17] it was only in the eighties that they called the attention of the community as a whole. The main reason for this is Unruh's seminal paper on "experimental black hole evaporation" [18]. Unruh conjectured that analogue models could help us to understand not only classical aspects of GR but also some features of quantum field theory in curved spacetimes. In particular, Unruh showed that an analogue black hole may present some properties of gravitational ones as far as the quantum thermal radiation is concerned. Since then, it became a common practice to use analog gravity as a tool to mimic different aspects of the gravitational theory: artificial black holes, Hawking radiation, emergent gravity, quantum field theory in curved backgrounds, and others.

In this paper, I adopt a different perspective concerning "artificial" gravity. I analyze if effective spacetime techniques are suitable to describe some aspects of Newtonian mechanics of point particles and vice versa. As it is known, the relation between curved geometries and particle dynamics was studied at least since the time of Jacobi [19]. After him, this geometrization scheme was to be the concern of Liouville, Lipschitz, Thomson, Tait, and Hertz (see [19, 20] for a detailed discussion). With the advent of tensor calculus, it became clear that there existed a map between the trajectories of certain mechanical systems in configuration space and the geodesics of a curved manifold. This fact was explored to solve mechanical problems of holonomic constraints by Ricci, Levi-Civita and in the relativistic era by Synge [21, 22], Lanczos [23], Lichnerowicz [24], and others. Nevertheless, despite the intense use of geometrical techniques in the context of dynamics, it seems that the relation between mechanics and geometry was not clearly appreciated in the literature of analog gravity (see, nevertheless, [25–27]). This is, perhaps, because many of the above maps did not take the time coordinate into account in the geometrization scheme.

It will be shown that Newton's mechanics of point particles in static potentials may provide a very simple analog model of gravitation. At first this statement seems to be suspicious once we will obtain an alternative description of Newtonian trajectories in terms of pseudo-Riemannian spacetimes without any reference to relativity theory. Nevertheless, using the well-known optical-mechanical analogy as a starting point we will see that, indeed, it is possible to give an effective spacetime description of the motion instead of using the traditional description in terms of forces. We will see that the resulting analog model works as a counterpart of Gordon's metric in the context of optics, while the trajectory of the particle is mapped into null geodesics of an effective spacetime with metric $\hat{g}_{\mu\nu}$. Finally, we explore some specific potentials and explicitly calculate the effective metrics.

This new geometrized scenario of mechanics may give us some interesting hints to the study of future models because of three main reasons. (1) It introduces curved spacetimes from a very simple and well understood physics. (2) It may extend our laboratory perspectives to measure effective gravitation using, for instance, electronic optics. (3) It may provide an interesting and smooth transition to the issue of quantization since we are working with the mechanics of particles instead of fields.

THE OPTICAL-MECHANICAL ANALOGY REVISITED

The well-known optical-mechanical analogy (OMA) has been discussed often and from many different points of view [25, 27]. The analogy may be understood as a formal tool which maps mechanical systems into optical ones and vice versa. In this sense it implies that the experience and insights developed in one area may be extrapolated to solve problems in the other. Although the OMA is quite familiar, I briefly discuss its main points to fix notation and to make the paper self-contained.

Let us start by considering a particle with mass "m" in a static potential $V(x^k)$. We are interested in the subset of particle trajectories with a fixed energy E_0 according to Newton's mechanics. It is instructive to begin with an arbitrary spatial coordinate system x^k $(k = 1, 2, 3)$ and three-dimensional euclidean space with metric $g_{ik}(x)$. Also, instead of using the usual time coordinate "t" as a parameter I use the arclength "l" along the curve, where

$$dl^2 = g_{ik}\dot{x}^i\dot{x}^k \, dt^2.$$

(1)

Defining the normalized tangent vector u^i to the curve it is straightforward to show that Newton's second law can be cast in the compact form

$$\left(pu^i\right)_{;k} u^k = g^{ij}\frac{\partial p}{\partial x^j},$$

(2)

where ";" is defined as the three-dimensional covariant derivative, and the modulus of the momentum is related to the potential by the usual energy relation

$$p = \sqrt{2m(E_0 - V)}.$$

(3)

It is important to note here that, because the modulus "p" is automatically known if x^k is given, we just need as initial conditions the position x_0^k and the direction of propagation u_0^k. If we specify only the initial position, the solutions of (2) characterize all possible trajectories passing through x_0^k with fixed momentum $p(x_0^k)$.

A similar situation appears in the context of geometrical optics. According to Fermat's principle see, for instance [28], the ray equation inside an heterogeneous isotropic material with index of refraction $n(x^k)$ is given by the extrema of the optical length $\lambda \equiv ct$

$$\lambda = \int n\,dl = \int n\sqrt{g_{ik}\dot{x}^i\dot{x}^k}\,dt,$$

(4)

where "c" is the velocity of light in vacuum. By varying the action and again parametrizing the paths with the arclength "l" it follows

$$\left(nu^i\right)_{;k} u^k = g^{ij}\frac{\partial n}{\partial x^j}.$$

(5)

In the same way, because the velocity of light is automatically given in terms of the index of refraction, one only needs to specify position and direction to obtain a given trajectory. By comparing (5) with (2) one immediately recognizes the similarity between the equations. Thus, there exists a clear analogy between Newtonian trajectories for point particles in static conservative fields and the light rays inside isotropic heterogeneous media.

To complete the analogy we divide (2) by $p_0 \equiv \sqrt{2m|E_0|}$ to express it in terms of a dimensional quantity $\hat{p} \equiv p/p_0$. Because the index of refraction is also an a dimensional quantity, one may define the map n

$$n \longmapsto \hat{p} = \sqrt{\pm\left(1 - \frac{V}{E_0}\right)},$$

(6)

where "+" stands if $E_0 > 0$ and "−" if $E_0 < 0$. In this way, the content of the OMA can be formulated as follows.

1.   To a given optical trajectory inside an isotropic material with index of refraction "n," it is possible to assign a static potential "V" given by relation (6) such that it mimics this trajectory in the context of Newtonian mechanics or vice versa.

It is important to note that the parallelism extends only to the geometrical form of the extremal curve. As was noted by Lanczos [23], the evolution of the systems in time is not equivalent. We will explore this fact later.

The Relation between Paths and Three-Dimensional Effective Geodesics

It is a well-known fact in optics that Fermat's principle may be mapped in the problem of finding geodesics of a three-dimensional curved manifold [23, 29]. If one interprets the optical length $d\lambda \equiv n dl$ as an infinitesimal element of arc, it is immediate to see that the extremal of the action (4) is completely equivalent to the geodesics of the riemannian geometry given by the metric

$$\widehat{g}_{ij} \equiv n^2 g_{ij}. \tag{7}$$

In other words, the light path is such that, if it is parametrized by the optical length λ it satisfies

$$\frac{d^2 x^k}{d\lambda^2} + \widehat{\Gamma}^k{}_{ij} \frac{dx^i}{d\lambda} \frac{dx^j}{d\lambda} = 0, \tag{8}$$

where the effective Christoffel symbol is such that

$$\widehat{\Gamma}^k{}_{ij} = \Gamma^k{}_{ij} + n^{-1} \left(\delta^k_i n_{,j} + \delta^k_j n_{,i} - n^{,k} g_{ij} \right). \tag{9}$$

Although this is a purely formal property of the action principle, one can convince himself/herself by explicitly substituting the identity $\lambda = n dl$ in (8) . We obtain, after a straightforward calculation and definition (9), the equation

$$n^2 \frac{d}{dl} \left(\frac{1}{n} u^k \right) + n \Gamma^k{}_{ij} u^i u^k + 2 n_{,j} u^j u^k = g^{kj} n_{,j}. \tag{10}$$

By calculating explicitly the first term one finally obtains

$$\frac{d}{dl} \left(n u^k \right) + n \Gamma^k{}_{ij} u^i u^k = g^{kj} n_{,j}, \tag{11}$$

which is no more than the ray equation (5) . The identification trajectories/ geodesics appears often in analog models of gravitation based on optics and is currently being explored in diverse contexts such as in the architecture of metamaterials, cloaking devices, negative refraction structures, and perfect magnifying lenses see (30–36) and references therein .

Less discussed in the literature of analogue gravity is the geometrical version of the mechanical counterpart of Fermat's principle. Nevertheless, the similarity between (2) and (5) is not fortuitous. Newton's equation as well is given in terms of an action principle of the form (4) which is Jacobi's principle. One only needs to replace n by $\widehat{P}$ in (4) to obtain (2). Thus, there exists also in mechanics a geometrical interpretation of the motion as described in its configuration space. The trajectories (2) may be alternatively obtained from the geodesics of the effective riemannian space given by the metric

$$\widehat{g}_{ij} = \widehat{p}^2 g_{ij}.$$

(12)

Note however that, although the four-dimensional generalization of (7) made by Gordon [17] became the paradigm of analog gravity, the same did not happen with its mechanical version. The aim of the next section is to explore this last possibility.

GORDON'S METRIC AND NEWTONIAN TRAJECTORIES

Gordon was the first to develop effective metric techniques in the context of analog models. He was interested in trying to describe dielectric media by an effective metric while at the same time using the gravitational field in Einstein ansatz to mimic a dielectric medium. Gordon showed that the trajectories of light inside a dielectric medium was such that they could be mapped in null geodesics of a four-dimensional pseudo-Riemannian metric given by

$$\widehat{g}_{\mu\nu} = \eta_{\mu\nu} + \left(\frac{c^{-2}}{\mu\epsilon} - 1\right)v_\mu v_\nu,$$

(13)

where v_μ is a normalized timelike vector with respect to Minkowski's background metric, that is $\eta^{\mu\nu}v_\mu v_\nu = 1$ and represents the motion of the dielectric medium. In the simplest case where the medium is at rest in laboratory's frame, we have $v^\mu = \delta^\mu_0$ and Gordon's metric acquire the simpler form

$$\widehat{g}_{\mu\nu} = \mathrm{diag}\left(n^{-2}, -g_{ij}\right).$$

(14)

Now, we are going to show that there exists an entirely equivalent situation in Newtonian mechanics of point paticles. We will see that Gordon's metric provides the apparatus to describe Newtonian motions by means of the optical-mechanical analogy. Although it is possible to give a formal demonstration of our proposition we are going to adopt a simpler route by showing that Newton's second law in the form (2) is completely equivalent to a null geodesic on the four-dimensional curved spacetime given by the metric (14).

We first note that the energy relation (3) can be put in the suggestive form

$$\widetilde{p}^{-2} - g_{ij}\widehat{p}^i\widehat{p}^j = 0,$$

(15)

where the first term has the index of refraction form given by (6) and $\widehat{}$ $\widehat{p}^i \equiv dx^i/d\lambda$. Our next aim is to write this equation as a quadratic form in a four-dimensional riemannian manifold. We thus define an effective spacetime $\widehat{M}$ with coordinates $x^\mu \equiv (\lambda, x^1, x^2, x^3)$ with λ given by (4) and (6) (note that the coordinate λ has the dimensions of length).

We are going to parametrize the trajectory of the particle in $\widehat{M}$ in terms of λ, that is, $x^\mu(\lambda)$. Defining the tangent four-vector $\widehat{p}^\mu$ as

$$\widehat{p}^\mu(\lambda) \equiv \frac{dx^\mu}{d\lambda},$$

(16)

we see that (15) is equivalent to the expression

$$\widehat{g}_{\mu\nu}\widehat{p}^\mu\widehat{p}^\nu = 0,$$

(17)

with the effective metric$\widehat{}$ $\widehat{g}_{\mu\nu}$ given by the Gordon metric (14) and the index of refraction given by the optical-mechanical relation (6). Relation (17) is the mechanical analogue of the light cone in relativity.

Note that, for the effective metric to make sense it has to carry a hyperbolic (Lorentzian) signature. This is in complete agreement with the fact that the modulus of the three-dimensional momentum "p" is real, and thus the potential is such that $(x) \leq E_0$. We thus succeeded to write the energy relation of Newtonian particles with fixed energy in terms of a "dispersion relation" in a pseudo-Riemannian geometry. Nevertheless, contrarily to our relativistic intuition, the trajectories of massive particles are not mapped into timelike geodesics, but are tangent to effective null curves.

To conclude our four-dimensional geometrization we are going to show that, indeed, the null geodesics of Gordon's metric coincide with Newton's equation in the form (2). To make our calculations simpler, we note that null geodesics are insensitive to conformal transformations. We thus make the conformal map by the substitution

$$\widehat{g}_{\mu\nu} \longmapsto \widehat{p}^2\widehat{g}_{\mu\nu},$$

(18)

to obtain a new manifold with metric given by

$$\hat{g}_{\mu\nu} = \text{diag}\left[1, \mp\left(1 - \frac{V}{E_0}\right)g_{ij}\right],$$

(19)

where "−" stands if $E_0 > 0$ and "+" if $E_0 < 0$. The four-dimensional geodesics of this geometry is given by

$$\frac{d^2 x^\mu}{d\lambda^2} + \hat{\Gamma}^\mu{}_{\alpha\beta}\frac{dx^\alpha}{d\lambda}\frac{dx^\beta}{\lambda} = 0.$$

(20)

We first note that, because the potential $V(x^k)$ depends only on position by hypothesis we have

$$\hat{\Gamma}^0{}_{\mu\nu} = \hat{\Gamma}^k{}_{\mu 0} = 0.$$

(21)

We thus obtain that, because $x_0 \equiv \lambda$, the "0" component of the equation is identically satisfied. Also, because we have $\hat{\Gamma}^k{}_{\mu 0} = 0$, the spatial components are immediately reduced to an equation of the form (8), which is identically satisfied as a consequence of Jacobi principle (one can convince herself/himself of this fact by explicitly substituting $d\lambda = \hat{p}v dt$ in the geodesic equation to obtain Newton's equations). We thus arrive at the following conclusion

The trajectories of particles with given energy E_0 in Newton's mechanics of static potentials are such that they can be derived from the null geodesics of a curved four-dimensional spacetime given by Gordon's metric up to a conformal transformation.

We thus define the effective four-dimensional element of distance as

$$d\hat{s}^2 \equiv d\lambda^2 \mp \left(1 - \frac{V}{E_0}\right)g_{ij}\, dx^i\, dxj.$$

(22)

Trajectories are simply given by the condition $d\hat{s}^2 = 0$. Note that the effective geometry is dependent on the total energy of the system. This means that we have to deal with different geometries to describe motions with different energies. Note also that, in general, the mechanical Riemann tensor $\hat{R}_{\alpha\beta\mu\nu}$ associated to the effective metric will be nonzero. At this point it is appropriate to use some words of Barcelo et al. in [15]: "It is quite remarkable that even though the underlying" particle "dynamics is Newtonian, nonrelativistic, and takes place in flat space plus time", the trajectories "are governed by a curved (3+1) dimensional Lorentzian (pseudo-Riemannian) spacetime geometry. For practitioners of general relativity this observation describes a very simple and concrete physical model for certain classes of Lorentzian spacetimes, including black holes. It is also potentially of interest to practitioners of particle "mechanics in that it provides a simple concrete introduction to Lorentzian differential geometric techniques."

EFFECTIVE GEOMETRIES FROM NEWTONIAN POTENTIALS

We now turn to some explicit situations where our geometrization procedure may be applied. Our aim is to show that some mechanical properties of the systems may as well be described in terms of an effective geometrical language.

The Geometry of Harmonic Oscillations

We start with a one-dimensional oscillator with potential energy $V = kx^2/2$. . The (1+1) dimensional effective line element (22) is of the form

$$d\hat{s}^2 \equiv d\lambda^2 - \left[1 - \left(\frac{x}{a}\right)^2\right]dx^2,$$

(23)

where "a" is the amplitude of the oscillator. Note that this metric is hyperbolic only if $|x|<a$. Thus, this effective spacetime is such that there exists two geometrical boundaries given by the conditions $x=\pm a$. This is an immediate consequence of the fact that we are considering only trajectories such that the total energy E_0 is fixed, and thus the motion is not defined anywhere.

The null trajectories may be obtained by imposing the condition $d\hat{s} = 0$, that is,

$$\frac{dx}{d\lambda} = \pm\left[1 - \left(\frac{x}{a}\right)^2\right]^{-1/2}.$$

(24)

In this simple case, it is possible to integrate this equation to obtain the implicit relation1

$$\lambda = \pm\frac{1}{2}\left\{x\sqrt{1 - \left(\frac{x}{a}\right)^2} + a\mathrm{Arcsin}\left(\frac{x}{a}\right)\right\} + \mathrm{const.}$$

(25)

A diagram for the trajectories is given in Figure 1. The set of all possible null curves (λ) given by the implicit relation (25) characterizes the space-time diagram for the oscillator. Note however that, although the curves exhibit a periodic behavior in terms of λ the derivative $dx/d\lambda$ is not well defined in the limit $x\rightarrow\pm a$. In this limit the effective volume $\sqrt{-\hat{g}}\,dx\,d\lambda = [1 - (x/a)]dx\,d\lambda$

also vanishes, and one can interpret this as a consequence of the fact that the effective spacetime only makes sense inside the boundaries given by $x=\pm a$.

Spacetime diagram

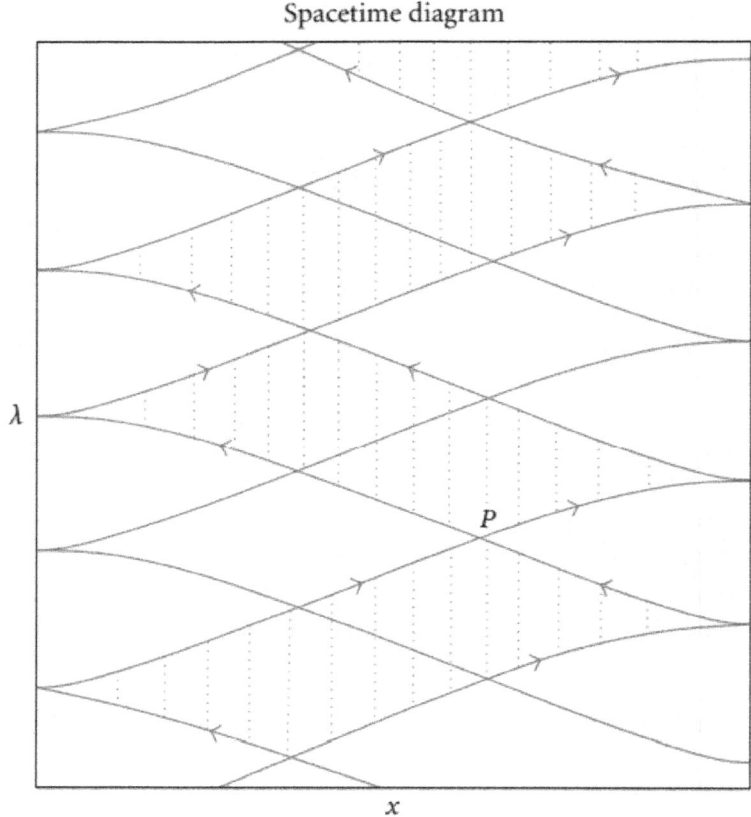

Figure 1: Harmonic oscillator space-time diagram.

At a given point of the diagram there exist only two admissible trajectories with fixed energy $E0$. These two curves determine at each point P a conoid structure in the same way as the propagation of light characterizes the light cone in relativity. We see from the figure that the intersection between the curves gives the mechanical analogue of the light cone at each spacetime point. The domain of causality of the point P is represented by the shaded regions.

The curves (25) are null geodesics with respect to the effective metric. Nevertheless, a careful inspection in (23) shows that this metric is flat. Thus, metric (23) is simply Minkowski metric in a curvilinear coordinate system. We are now going to investigate a more complicated situation where the Riemann tensor does not vanish.

The Geometry of Newtonian Gravitation

We now turn to the motion of point test particles with mass "m" in the Newtonian gravitational potential. As in the previous section, we are interested in the motions associated to a given total energy E_0. According to our previous discussion, the null geodesics of the respective effective metric (22) are such that they reproduce ordinary Newtonian trajectories in the gravitational potential.

To be specific, let us consider the spherically symmetric potential $V(r) = -GMm/r$. We will consider only the case of bounded orbits, that is, $E_0 < 0$. The other situations will be considered elsewhere.

$E_0 < 0$ Effective Metrics

In this case we have a maximum admissible radii $r_H \equiv -GMm/E_0$ for the trajectories. One obtains for the effective metric, in spherical coordinates, the expression

$$d\hat{s}^2 = d\lambda^2 + \left(1 - \frac{r_H}{r}\right)\left[dr^2 + r^2 d\Omega^2\right].$$

(26)

Analogously to the harmonic oscillator case, we obtain an effective metric that admits a hyperbolic (Lorentzian) signature only in a particular domain. In the present case this domain is given by the condition $0 < r < rH$. This is because outside this region, that is, for $r > rH$, the movement is not allowed for the considered energies $E_0 = -GMm/r_H$.

This fact has an interesting geometrical interpretation in our scheme. First, let us note that the square root of the determinant is given by

$$\sqrt{-\hat{g}} = r^2 \sin(\theta)\left(\frac{r_H}{r} - 1\right)^{3/2}.$$

(27)

This relation implies that the element of effective volume $dV_{\text{eff}} \equiv \sqrt{-\hat{g}}\, d^4x$ vanishes for the values $r=0$ and $r=rH$. This is a hint that the confinement of the motion between these values of the radial coordinate has a deep geometrical meaning.

In fact, the components of the Riemann tensor are

$$\hat{R}_{1212} = -\frac{1}{2}\frac{r_H}{r_H - r},$$

$$\hat{R}_{1313} = -\frac{1}{2}\frac{r_H\sin^2(\theta)}{r_H - r},$$

$$\hat{R}_{2323} = -\frac{1}{4}\frac{rr_H\sin^2(\theta)(3r_H - 4r)}{r_H - r}.$$

(28)

One immediately obtains for the Ricci scalar the expression

$$R = -\frac{3}{2}\frac{r_H^2}{r(r_H - r)^3}.$$

(29)

This object diverges for the values $r=0$ and $r=rH$ (note that $R<0$). Thus, the effective (3+1) spacetime is delimited by two singularities in the effective geometry. The first one is quite similar to Schwarzchild singularity and admits a similar interpretation due to the strength of the gravitational field. It appears because the Newtonian gravitational potential also diverges at the origin. The second singularity has, nevertheless, a different and unexpected origin. It appears because we confined our attention only to the movement of particles with a fixed energy E_0. As we mentioned before, all possible trajectories with this energy cannot scape from the region $r<rH$. This fact appears in the geometrical description as a kind of geometrical barrier that do not allow the null geodesics to escape from the allowed region.

To develop some additional feeling on our geometrical description of dynamics, it is instructive to solve the well-known problem of finding the trajectories in the Newtonian gravitational potential in terms of the effective null geodesics (20). We recall that the following quantity is a constant of the motion:

$g\mu v dx\mu d\lambda dxv d\lambda = 0$. (30)

From the geodesic equation we obtain (for the sake of conciseness I use $dx/d\lambda \equiv \dot{x}$)

$$\ddot{r} + A(r)\dot{r}^2 + B(r)\left[\dot{\theta}^2 + \sin^2\theta\dot{\phi}^2\right] = 0,$$

(31)

$$\ddot{\theta} + C(r)\dot{r}\dot{\theta} - \sin\theta\cos\theta\dot{\phi}^2 = 0,$$

(32)

$$\ddot{\phi} + C(r)\dot{\phi}\dot{r} + 2\cot g\theta\dot{\phi}\dot{\theta} = 0,$$

(33)

Where

$$A(r) \equiv \frac{1}{2} \frac{r_H/r}{r - r_H},$$

$$B(r) \equiv -\frac{1}{2} \frac{2r^2 - r_H r}{r - r_H},$$

$$C(r) \equiv \frac{2r - r_H}{r^2 - r_H r}.$$

(34)

As usual, we assume that the orbit lies in the plane $\theta = \pi/2$. Thus, (32) ensures that, if $\theta = \pi/2$ and $\dot\theta = 0$ initially, the orbit remains always in this plane. Equation (33) may be integrated directly. We obtain that the following quantity is also conserved

$$\left(r^2 - r_H r\right)\dot\phi = \xi,$$

(35)

where ξ is a constant. It is straightforward to show that this equation implies the usual conservation of angular momentum. In fact, from (35), it follows, after a reparametrization that

$$\frac{d\phi}{dt} = \frac{l}{r^2}, \quad l = \left(-\frac{2E_0}{m}\right)^{1/2}\xi.$$

(36)

We finally turn to (31). As expected, this is not a simple equation. Nevertheless, we can bypass this difficulty by considering the constraint (30). We obtain

$$\dot r^2 = -\frac{1}{(1 - r_H/r)}\left[1 + \frac{(\xi/r)^2}{(1 - r_H/r)}\right].$$

(37)

This is the equation of motion of the radial part written in terms of the parameter λ. By introducing the auxiliary variable l

$$u = \frac{1}{r},$$

(38)

we obtain

$$\xi^2\left(\frac{du}{d\phi}\right)^2 = -\left(1 - r_H u + \xi^2 u^2\right).$$

(39)

Deriving with respect to ϕ we arrive at the familiar Newtonian equation for the orbit

$$\frac{d^2u}{d\phi^2} + u - \frac{GM}{l^2} = 0,$$

(40)

which admits the solution*l*

$$r(\phi) = \frac{l^2/GM}{1 + \epsilon\cos(\phi - \phi_0)},$$

(41)

where $0\leq\epsilon<1$ because $r\leq rH$ always. Thus, we succeeded to find the orbits around the field of a sphericall object with mass "M" in terms of the null geodesics of the effective metric (26).

The Geometry of Electrostatic Potentials

The conservative potentials that appear in electromagnetism are, perhaps, the most interesting from the point of view of analog gravity in our context. This is not only because electromagnetic fields are simple to handle in laboratory sizes, but because many interesting configurations of static potentials occur in the advanced field of electronic optics. I will concentrate here on simple electrostatic configurations. The potential is obtained by the familiar formula

$$\phi(\vec{x}) = \frac{1}{4\pi\epsilon_0}\int\frac{\rho(\vec{x}')}{|\vec{x} - \vec{x}'|}d^3x',$$

(42)

while the potential energy is $V(\vec{x}) = q\phi$. The effective metric is given by the expression

$$\hat{g}_{\mu\nu} = \text{diag}\left[1, \mp\left(1 - \frac{q\phi}{E_0}\right)g_{ij}\right],$$

(43)

where q is the charge of the test charge. Our next task is to calculate this metric for a specific case and to use some of its geometrical properties to infer about mechanical properties of the system.

The Case of the Electric Dipole

One interesting situation appears in the geometrization of the static electric dipole. If the dipole moment $\vec{p}$ is oriented in the z direction, we obtain, for large distances,

$$\phi(r,\theta) = \frac{p}{4\pi\epsilon_0}\frac{\cos(\theta)}{r^2}.$$

(44)

For the sake of simplicity lets us consider $pq>0$ and positive energies $E0>0$. The effective line element reads

$$d\hat{s}^2 = d\lambda^2 + \left(\kappa^2\frac{\cos(\theta)}{r^2} - 1\right)\cdot\left[dr^2 + r^2 d\Omega^2\right],$$

(45)

where $\kappa^2 \equiv pq/4\pi\epsilon_0 E_0$. Note that this effective spacetime only makes sense if

$$r^2 > \kappa^2\cos(\theta).$$

(46)

This is an immediate consequence of the fact that the energy $E0$ is fixed from the beginning and there are regions that are not allowed for the particle motion. In our scheme, this fact has an interesting geometrical interpretation. Note that the determinant of the metric

$$\sqrt{-\hat{g}} = r^2\sin(\theta)\left(1 - \frac{\kappa^2\cos(\theta)}{r^2}\right)^{3/2}$$

(47)

vanishes for some values of the radii. At first we could suspect that the coordinate system we are using is not well defined in this region. Nevertheless, we obtain for the scalar of curvature the expression

$$R = \frac{3\kappa^4}{2}\frac{\left(3\cos(\theta)^2 + 1\right)}{(\kappa^2\cos(\theta) - r^2)^3},$$

(48)

which diverges in the abovementioned region. Thus, the description of charged particle motions (such as electrons or ions) in terms of null geodesics of an effective spacetime forces us to deal with a singular geometry in the case of the electric dipole. Nevertheless, the singularity is not so simple as it appears in Schwarzschild geometry, but it is determined by a compact bidimensional region given by the condition

$$r^2 = \kappa^2\cos(\theta).$$

(49)

The null geodesics are not defined in the interior region. Note, however, that the entire region outside the singularity is well defined. Nevertheless, the geodesics are not so simple to be integrated as in the case of the solar system. Also, it is immediate to see that this geometry is asymptotically flat. Thus, for very large values of r, where the field of the dipole tends to vanish, the trajectories are described by the null geodesics of Minkowski spacetime.

As a final remark, I would like to point out that the geometrical framework may introduce techniques of general relativity and riemannian geometry in the realm of electron optics. We suspect that it is possible to envisage situations where electronic dispositives may be projected using these new tools. This last statement is strongly suggested by the recent achievements of transformation optics [31–35].

CONCLUSIONS

The relation between geometry and dynamics is an interesting problem that has been investigated from many different perspectives along the years. In this paper it is shown that Newtonian mechanics of point particles in static potentials admits a geometrical interpretation in terms of (3+1) effective spacetimes in a similar way as it appears in the context of analog gravity models. The mechanical effective metric $\hat{g}\mu\nu$ is given explicitly in terms of the potential energy (x) and the total energy $E0$ and is similar to Gordon's metric in optics. From this last similarity, we inferred that mechanical systems may provide a very simple arena for the study of analogue gravity. This new arena may bring interesting theoretical and experimental challenges. I investigated some explicit examples, where the use of geometrical techniques were used to infer about some mechanical properties. Thus, it may be natural to speak about effective horizons, singularities and other structures that appear in general relativity in mechanical problems also. In particular it was considered the motion of particles in the gravitational field of a central mass and in the potential of an electric dipole. The null geodesics of the above geometries coincides with the usual trajectories as they're described in the Newtonian ansatz. It may be interesting to investigate if these geometrical picture can be sustained after first quantization.

ACKNOWLEDGMENTS

The author would like to thank M. Novello, F. T. Falciano, S. E. P. Bergliaffa, and M. Borba for usefull comments. This work is supported by Faperj, Brazil.

REFERENCES

1.   M. Novello and E. Goulart, "Beyond analog gravity: the case of exceptional dynamics," Classical and Quantum Gravity, vol. 28, no. 14, Article ID 145022, 2011.

2. G. Krein, G. Menezes, and N. F. Svaiter, "Analog model for quantum gravity effects: phonons in random fluids," Physical Review Letters, vol. 105, Article ID 131301, 2010.

3. M. Visser and C. Molina-Paris, "Acoustic geometry for general relativistic barotropic irrotational fluid flow," New Journal of Physics, vol. 12, Article ID 095014, 2010.

4. C. Barcelo, S. Liberati, and M. Visser, "Refringence, field theory and normal modes," Classical and Quantum Gravity, vol. 19, no. 11, pp. 2961–2982, 2002.

5. M. Visser, C. Barcelo, and S. Liberati, "Bi-refringence versus bi-metricity," http://arxiv.org/abs/gr-qc/0204017v1.

6. T. A. Jacobson and G. E. Volovik, "Event horizons and ergoregions in 3He," Physical Review D, vol. 58, 7 pages, 1998.

7. R. Schutzhold, "Emergent horizons in the laboratory," Classical and Quantum Gravity, vol. 25, Article ID 114011, 2008.

8. M. Novello, "Effective geometry in nonlinear electrodynamics," International Journal of Modern Physics A, vol. 17, no. 29, pp. 4187–4196, 2002.

9. M. Novello and J. M. Salin, "Nonlinear electrodynamics can generate a closed spacelike path for photons," Physical Review D, vol. 63, Article ID 103516, 5 pages, 2001.

10. M. Novello and J. M. Salim, "Effective electromagnetic geometry," Physical Review D, vol. 63, Article ID 083511, 4 pages, 2001.

11. M. Novello, V. A. De Lorenci, J. M. Salim, and R. Klippert, "Geometrical aspects of light propagation in nonlinear electrodynamics," Physical Review D, vol. 61, no. 4, 10 pages, 2000.

12. C. Mayoral, A. Recati, A. Fabbri, R. Parentani, R. Balbinot, and I. Carusotto, "Acoustic white holes in flowing atomic Bose-Einstein condensates," New Journal of Physics, vol. 13, Article ID 025007, 2011. ·

13. U. Leonhardt and T. G. Philbin, "Chapter 2 transformation optics and the geometry of light," Progress in Optics, vol. 53, pp. 69–152, 2009.

14. S. Weinfurtner, E. W. Tedford, M. C. J. Penrice, W. G. Unruh, and G. A. Lawrence, "Measurement of stimulated hawking emission in an analogue system," Physical Review Letters, vol. 106, 4 pages, 2011. ·

15. C. Barcelo, S. Liberati, and M. Visser, "Analogue gravity," Living Reviews in Relativity, vol. 8, no. 12, 2005.

16. M. Novello, M. Visser, and G. Volovik, Eds., Artificial Black Holes, World Scientific, 2002.

17. W. Gordon, "Zur lichtfortp anzung nach der relativitat-stheorie," Annalen der Physik, vol. 72, pp. 421–456, 1923.

18. W. G. Unruh, "Experimental black-hole evaporation?" Physical Review Letters, vol. 46, no. 21, pp. 1351–1353, 1981.

19. R. Dugas, A History of Mechanics, Dover Publications, 1988.

20. J. Lutzen, Mechanistic Images in Geometric Form, Oxford University Press, 2005.

21. J. L. Synge, "On the geometry of dynamics," Transactions of the Royal Society, 1926.

22. J. L. Synge, "Mechanical Models of Spaces With Positive-Definite Line-Element," Annals of Mathematics, vol. 36, no. 3, 1935.

23. C. Lanczos, The Variational Principles of Mechanics, Dover Publications, 4th edition, 1986.

24. A. Lichnerowicz, "Sur la transformation des equations de la dynamique," Comptes rendus de l›Académie des Science de Paris, vol. 223, pp. 649–651, 1946.

25. J. Evans and M. Rosenquist, American Journal of Physics, vol. 54, no. 10, p. 876, 1986.

26. J. Evans, K. K. Nandi, and A. Islam, "The optical-mechanical analogy in general relativity: new methods for the paths of light and of the planets," American Journal of Physics, vol. 64, no. 11, pp. 1404–1415, 1996.

27. J. Evans, K. K. Nandi, and A. Islam, "The optical-mechanical analogy in general relativity: exact Newtonian forms for the equations of motion of particles and photons," General Relativity and Gravitation, vol. 28, no. 4, pp. 413–439, 1996.

28. J. L. Synge, Geometrical Optics: An Introduction to Hamilton›s Method, Cambridge University Press, 1937.

29. A. Lichnerowicz, Ed., Elements de Calcul Tensoriel, Jacques Gabay, 1987.

30. D. A. Genov, S. Zhang, and X. Zhang, "Mimicking celestial mechanics in metamaterials," Nature, vol. 5, no. 9, pp. 687–692, 2009.

31. U. Leonhardt and T. G. Philbin, "General relativity in electrical engineering," New Journal of Physics, vol. 8, article 247, 2006.

32. U. Leonhardt and T. Tyc, "Broadband invisibility by non-euclidean cloaking," Science, vol. 323, no. 5910, pp. 110–112, 2009.

33. V. M. Shalaev, "Physics: transforming light," Science, vol. 322, no. 5900, pp. 384–386, 2008.

34. D. Schurig, J. J. Mock, B. J. Justice et al., "Metamaterial electromagnetic cloak at microwave frequencies,"Science, vol. 314, no. 5801, pp. 977–980, 2006.

35. R. A. Shelby, D. R. Smith, and S. Schultz, "Experimental verification of a negative index of refraction,"Science, vol. 292, no. 5514, pp. 77–79, 2001.

36. O. Hess, "Optics: farewell to flatland," Nature, vol. 455, no. 7211, pp. 299–300, 2008.

Chapter 9

A HYBRID APPROACH FOR THE RANDOM DYNAMICS OF UNCERTAIN SYSTEMS UNDER STOCHASTIC LOADING

Michele Betti, Paolo Biagini, and Luca Facchini

Department of Civil and Environmental Engineering (DICeA), University of Florence, Street Santa Marta 3, 50139 Florence, Italy

ABSTRACT

This paper presents a hybrid Galerkin/perturbation approach based on Radial Basis Functions for the dynamic analysis of mechanical systems affected by randomness both in their parameters and loads. In specialized literature various procedures are nowadays available to evaluate the response statistics of such systems, but sometimes a choice has to be made between simpler methods (that could provide unreliable solutions) and more complex methods (where accurate solutions are provided by means of a heavy computational effort). The proposed method combines a Radial Basis Functions (RBF) based Galerkin method with a perturbation approach for the approximation of the system response. In order to keep the number of differential equations to be solved as low as possible, a Karhunen-Loève (KL) expansion for the excitation is used. As case study a non-linear single degree of freedom (SDOF) system with random parameters subjected to a stochastic windtype load is analyzed and discussed in detail; obtained numerical solutions are compared with the results given by Monte Carlo Simulation (MCS) to provide a validation of the proposed approach. The proposed method could be a valid alternative to the classical procedures as it is able to provide satisfactory approximations of the system response.

INTRODUCTION

The analysis of systems with uncertain parameters is an increasingly important subject in the treatment of a plethora of engineering problems. The dynamic response of a mechanical system with uncertain parameters possesses probabilistic features which depend on the probability distribution of the system parameters, and uncertainty in geometry or mechanical properties could affect both response and reliability of a structure [1, 2]. It is then necessary to take into account the consequence of these uncertainties in the study and design as randomness of structural parameters may lead to large and unexpected excursions of the structural response with, in some cases, drastic reductions in structural reliability. The reliability analysis for such structures strongly depends on the variation of these parameters; consequently, a probabilistic approach is necessary for adequate reliability analysis. Uncertainties must be adequately described to be included in a mathematical model that can be conveniently solved. To this aim, probability theory has been extensively adopted as a general tool able to characterize structural uncertainty by means of random fields and random variables [3].

The most common technique employed to analyse systems with uncertain parameters is the Monte Carlo simulation (MCS). This technique allows the statistical evaluation of the system response based on a certain number of analyses with different values of the random parameters. This approach could be very computationally expensive since it requires, after generating a large number of samples of uncertain parameters, the solution of the corresponding deterministic (linear or nonlinear) problem. Random numbers, representing uncertain parameters with given correlation structure, are usually generated by digital simulation [4]. The main advantage of the MCS method is that the procedures adopted in a deterministic setting can be directly applied to solve the mechanical problem at hand. In general, however, especially for nonlinear systems, the simulation procedures are quite inefficient due to the large number of samples needed to guarantee accurate statistical results. Moreover, the implementation of MCS requires the complete probabilistic description of the uncertain quantities through their probability density functions, which are often unavailable. All these remarks make evident that the Monte Carlo-based structural analysis, even if still the most used among the stochastic analysis methods for structural problems, may become very heavy from a computational point of view as the number of structure degrees of freedom (DOFs) and the number of uncertain parameters increase. In some cases, the computational effort makes it inapplicable (especially in case of nonlinearities), so the MCS method is mainly used as a test for analytical approaches.

For these reasons, alternative procedures have been proposed in the literature. Methods based on perturbation of the stochastic quantities have been presented both for linear [5, 6] and nonlinear problems and have been applied in the solutions of several engineering problems, such as frame analysis, reliability analysis, and buckling of shells with random initial imperfections [7, 8]. These approaches employ the conventional finite element method along with perturbation techniques and allow the probabilistic characterization of the response in terms of mean and covariance functions once the first- and second-order moments of the uncertainties are assigned. These methods mainly consist of a direct approach using probabilistic, instead of statistic, theory. This is usually pursued both in static and dynamic settings by using, for instance, expansion methods, where the stiffness matrix of the structural problem is split into a deterministic part (obtained with the mean value of random parameters) and into a part which accounts for the fluctuation of the random variables about its mean value. In order to evaluate the probabilistic response, Taylor expansions or Neumann expansions [9–11] are adopted to avoid inversion of matrices depending on the random parameters. Most of them, as pointed out before, are based on perturbation techniques, so that the stochastic finite element (SFE) method is often identified as the classical finite element (FE) method coupled with a perturbation approach [12,13]. Unfortunately, some of these approaches show a drawback in that they are less and less accurate as the level of the uncertainty of the parameters increases. Moreover, even if the uncertainty is low, they ensure accuracy only for the second-order statistics of the response. Consequently, they can be applied only in case of Gaussianity of the response, which is rarely the case, even if the basic uncertain parameters are modelled as Gaussian. In fact, due to the nonlinear relationship between the system response and the system (random) parameters, the response is usually strongly non-Gaussian, even for linear systems.

To overcome these drawbacks, hybrid approaches have been proposed in the past. One of these has been proposed by Ghanem [14] by coupling the Monte Carlo simulation and the spectral stochastic finite element method (SSFEM). The author proposes to use a polynomial chaos expansion of the system response that, after substitution in the original equilibrium equation, leads to residuals that are orthogonalised with respect to the expansion basis. The original problem is then reduced to an iterative solution of a linear system of equations. Nevertheless, even if published in late 1998, no convincing application of these ideas has been published so far, and the methods seem practically limited to linear problems. A method due to Liu et al. [9] shows how an estimation of the time history of first two moments for the structural response in a linear or nonlinear system is obtained. This work has been improved by Chiostrini and Facchini [15] where a stochastic input has also been taken into account. The first two moments of the response have been evaluated taking into account a Taylor expansion of the structural response centred on the mean value of random parameters. The method is efficient when the dependence between the response and the random parameters is approximately a polynomial of the same degree as the expansion. It is then necessary to take in account more flexible expressions for the approximation of the response.

In the present paper, to provide an insight into the statistical response variations of nonlinear systems, a computationally efficient approach is proposed that takes into account randomness both in mechanical parameters and loads. The approximation of the system response is accomplished by means of an expansion of radial basis functions (RBFs): a novel class of approximating functions is adopted to model the dependence of the response displacements upon the uncertain variables. Such expansion can be classified among the so-called radial basis neural networks with time-dependent coefficients [16, 17]. A stochastic process which can be described by this model allows a direct evaluation of its main characteristics leading to a remarkable accuracy level for the statistics of the response and, in general, for the probability density function (PDF). First, after formulating the problem, the approach is briefly drafted with the aim of introducing some useful quantities and notations; then, it is particularized with reference to the system under examination: a strongly nonlinear SDOF system with an uncertain parameter subject to a random load. The numerical treatment of the mechanical problem is discussed in detail. In particular, to develop the approach, two methods are mainly taken into consideration in the present work: the first is described in Liu et al. [9], and subsequently enhanced by Chiostrini and Facchini [15]. It can be classified as a perturbation method and makes use of sensitivity vectors to evaluate the first two moments of the response. As already mentioned, the efficiency of this method is closely linked to the degree of the expansion and of the dependence of the response on the random parameters. This condition is difficult to check, and this fact can lead to great errors in the results. The second one makes use of a Galerkin approach with radial basis trial functions: it is more complex and leads to the solution of a great number of differential equations than the former, but the results are much more satisfactory (Facchini [18] and Betti et al. [4]). Both methods are briefly reviewed for convenience in the following sections. The proposed procedure can be seen as a compromise between such different situations: it provides a satisfactorily accurate solution by means of not so complex calculations. As shown in the numerical application, the features of the improved perturbation method are preserved even in the nonlinear dynamic analysis. The results highlight a good level of accuracy as is revealed by comparisons with the MCS results. As it provides a good compromise between computational cost and accuracy, the proposed hybrid approach could be an effective tool for the solution of a wide range of problems involving uncertain parameters.

THEORETICAL REMARKS

The mechanical system considered in this study is assumed to be described by an equation of motion where a nonlinear restoring function characterizes the nonlinear structural behaviour; load is assumed as a random process:

$$\mathbf{M}\ddot{\mathbf{x}}+(\mathbf{x},\dot{\mathbf{x}},\mathbf{b})=\mathbf{F}(t) \tag{2.1}$$

In (2.1), $\mathbf{x}$ is the vector of the system degrees of freedom (DOFs), $\mathbf{M}$ is the mass matrix, and $\mathbf{g}$ is the nonlinear restoring function; dot indicates differentiation with respect to time. The nonlinear function $\mathbf{g}$ is assumed to be affected by randomness, that is, it changes according to the variability of some random parameters (e.g., Young's modulus of the structural material, dimensions of the structural members, etc.). Such uncertain parameters are grouped in the vector $\mathbf{b}$. The excitation (t) is a random process which can be described in terms of a parametric model by means of its mean vector $\mathbf{F}_m$ and some zero-mean random vectors $\boldsymbol{\alpha}^{(k)}$ with proper covariance structure:

$$\mathbf{F}(t) = \mathbf{F}_m + \sum_k \boldsymbol{\alpha}^{(k)}\chi_k(t). \tag{2.2}$$

Eventually, $\chi(t)$ denotes a function of time to be selected by the user.

The probabilistic properties of the response function $\mathbf{x}(t)$ can thus be obtained by evaluating the dependence, at each instant, of the response function itself with respect to the random parameters (i.e., the analysis of the function $\mathbf{x}(\mathbf{b},\boldsymbol{\alpha},t)$). This analysis, as discussed in the introduction, can be approached by several methods. Two of them are, for convenience, next reported: the perturbation method and the Galerkin approach. In particular, the Galerkin method is next reviewed considering an expansion of radial basis functions (RBFs).

Brief Review of Perturbation Method

The perturbation method approximates the dependency of the system response on the random vectors $^{(k)}$ and $\mathbf{b}$, under the assumption of their independency, by means of a series expansion:

$$x_k\left(t,\mathbf{b},\boldsymbol{\alpha}^{(h)}\right) = x_k(t,\mu_\mathbf{b},\mu_\alpha) + \frac{\partial\left[x_k(t,\mu_\mathbf{b},\mu_\alpha)\right]}{\partial\mathbf{b}}(\mathbf{b}-\mu_\mathbf{b}) + \frac{\partial\left[x_k(t,\mu_\mathbf{b},\mu_\alpha)\right]}{\partial\boldsymbol{\alpha}^{(i)}}\left(\boldsymbol{\alpha}^{(i)}-\mu_{\alpha^{(i)}}\right)$$

$$+ \frac{1}{2}(\mathbf{b}-\mu_\mathbf{b})^t\frac{\partial^2\left[x_k(t,\mu_\mathbf{b},\mu_\alpha)\right]}{\partial\mathbf{b}^2}(\mathbf{b}-\mu_\mathbf{b})$$

$$+ \frac{1}{2}\left(\boldsymbol{\alpha}^{(i)}-\mu_{\alpha^{(i)}}\right)^t\frac{\partial^2\left[x_k(t,\mu_\mathbf{b},\mu_\alpha)\right]}{\partial\boldsymbol{\alpha}^{(i)}\partial\boldsymbol{\alpha}^{(j)}}\left(\boldsymbol{\alpha}^{(j)}-\mu_{\alpha^{(j)}}\right), \tag{2.3}$$

where μ is the mean of the random parameters. A repeated index implies summation over it. The response statistics can thus be obtained taking expected values of (2.3). The derivatives of the system response with respect to the random coefficients can be evaluated differentiating the equation of motion (2.1) (see e.g., Liu et al. [9] and Chiostrini and Facchini [15] for details). The method provides effective results when the dependence of the response on the random parameters is quadratic; a severe error can occur if this dependence is more complex (see [19]).

Brief Review of Galerkin-RBF Approach

The Galerkin approach describes the dependence of the response on the random parameters by means of time-dependent vectors $^{(k)}(t)$ and trial functions $\varphi_k(\mathbf{b}, \boldsymbol{\alpha}^{(h)})$:

$$x\left(t, \mathbf{b}, \boldsymbol{\alpha}^{(h)}\right) \cong \tilde{x}\left(t, \mathbf{b}, \boldsymbol{\alpha}^{(h)}\right) = \sum_{k=1}^{N_\varphi} w^{(k)}(t) \cdot \varphi_k\left(\mathbf{b}, \boldsymbol{\alpha}^{(h)}\right),$$

(2.4)

where N_φ is the number (chosen by the user) of the trial functions. A variational approach (Facchini [18] and Betti et al. [4]) can be established, to obtain the solution. Denoting by the symbol [·] the expected value operator, the equation that allows to obtain the solution can be written as follows:

$$E\left[\mathbf{e}\varphi_k\left(\mathbf{b}, \boldsymbol{\alpha}^{(h)}\right)\right] = 0 \quad \forall k = 1 \cdots N_\varphi,$$

(2.5)

where the error function $\mathbf{e}$ is defined as follows:

$$\mathbf{e}(t) = \mathbf{M}\tilde{\tilde{x}} + g\left(\tilde{x}, \tilde{\tilde{x}}, t\right) - \mathbf{F}(t).$$

(2.6)

In the classical Galerkin approach, the trial functions $\varphi(\mathbf{b}, \boldsymbol{\alpha}^{(h)})$ in (2.4) are fixed, and therefore, the solution is completely defined by (2.5). On the other hand, radial basis functions (RBFs) are dependent on some more parameters (the centre and the decay coefficients), which enable a further optimization of the procedure. A number of different RBFs can be found in the literature (see f.i. Broomhead and Lowe [16], Haykin [20], Poggio and Girosi [17], Gotovac and Kozulic [21], Mai-Duy and Tran-Cong [22]). In the present study, Gaussian bell-shaped RBF will be used. Their form is as follows:

$$\varphi_k\left(\mathbf{b}, \boldsymbol{\alpha}^{(h)}\right) = \phi\left(\frac{\|\mathbf{v} - \mathbf{c}^{(k)}\|}{\sigma_k}\right); \quad \mathbf{v} = \left[\mathbf{b}\boldsymbol{\alpha}^{(1)} \cdots \boldsymbol{\alpha}^{(n)}\right]^t,$$

(2.7)

With

$$\phi(r) = \exp\left(-r^2\right),$$

(2.8)

where $\mathbf{c}^{(k)}$ is the centre of the kth function and σ_k is its decaying parameter. It is to be noted that other kinds of RB functions work similarly well, as all of them are universal approximators [21]. In other words, under general conditions, arbitrarily complex functions can be approximated to a desired tolerance with a sufficiently high number of RBF with proper parameters [18, 23]. The variability of the centres and the decay parameters of the RBFs induces a variability of the RBFs themselves, so that another set of equations can be derived from the variational approach and an optimization procedure can be performed (see Facchini [18] for details).

The Proposed Hybrid Approach

The main idea of the approach herein proposed is to combine the two previously cited methods to join their advantages. Extremely summarizing the perturbation approach, from a numerical point of view, requires both a smaller amount of memory and a computational effort. The Galerkin method is more accurate and its accuracy can be controlled by means of error functions. Moreover, the perturbation method gives satisfactory results when it is used to model the dependence of the response on the random parameters of the forcing process, while modelling the dependence of the response on the random parameters of the mechanical system generally requires the use of the Galerkin approach [24]. Consequently the hybrid approach aims to combine the advantages of the two methods modelling the two different types of dependency in the system equation. Based on this idea the forcing process and the structural response are expressed as follows:

$$F(t) = \sum_{h=1}^{N_F} \boldsymbol{\alpha}^{(h)} a_h(t),$$

(2.9

$$x\left(t, \mathbf{b}, \boldsymbol{\alpha}^{(h)}\right) \cong \tilde{x}\left(t, \mathbf{b}, \boldsymbol{\alpha}^{(h)}\right) = \sum_{k=1}^{N_w} \mathbf{w}^{(k)}\left(t, \boldsymbol{\alpha}^{(h)}\right) \varphi_k(\mathbf{b}).$$

(2.10)

Eventually, an expansion like (2.3) is used:

$$\mathbf{w}^{(k)}\left(t, \boldsymbol{\alpha}^{(h)}\right) = \mathbf{w}^{(k)}\left(t, \mu_{\alpha}\right) + \frac{\partial \mathbf{w}^{(k)}\left(t, \mu_{\alpha}\right)}{\partial \boldsymbol{\alpha}^{(h)}}\left(\boldsymbol{\alpha}^{(h)} - \mu_{\alpha^{(h)}}\right)$$

$$+ \frac{1}{2}\left(\boldsymbol{\alpha}^{(m)} - \mu_{\alpha^{(m)}}\right)\frac{\partial^2 \mathbf{w}^{(k)}\left(t, \mu_{\alpha}\right)}{\partial \boldsymbol{\alpha}^{(m)}\partial \boldsymbol{\alpha}^{(h)}}\left(\boldsymbol{\alpha}^{(h)} - \mu_{\alpha^{(h)}}\right).$$

(2.11)

Then, the solution of original problem of (2.1) can be solved as a set of coupled deterministic nonlinear equations, and the statistics of the response process (mean value and correlation function) can be directly evaluated by means of the following:

$$E[\mathbf{x}(t)] = \sum_{k=1}^{N_w} E\left[\mathbf{w}^{(k)}(t, \boldsymbol{\alpha})\right] E\left[\varphi_k(\mathbf{b})\right],$$

$$\mathbf{R}_{XX}(t, s) = E\left[\mathbf{w}^{(h)}(t, \boldsymbol{\alpha})\mathbf{w}^{(k)}(s, \boldsymbol{\alpha})\right] E\left[\varphi_h(\mathbf{b})\varphi_k(\mathbf{b})\right]$$

$$- E\left[\mathbf{w}^{(h)}(t, \boldsymbol{\alpha})\right] E\left[\mathbf{w}^{(k)}(s, \boldsymbol{\alpha})\right] E\left[\varphi_h(\mathbf{b})\right] E\left[\varphi_k(\mathbf{b})\right].$$

(2.12)

Specific details of the proposed hybrid approach are reported next, where the method is applied to illustrative case studies.

NUMERICAL EXAMPLES

To explain and discuss feasibility of the proposed hybrid approach, several case studies are considered. As a first step, the deterministic loading has been considered with the aim to compare the result of the approach with the existing method; next, the case with stochastic loading is reported. The cases are discussed detailing each step of the proposed approach.

Deterministic Loading

Duffing Oscillator

As a first application, a SDOF Duffing oscillator subjected to a time-dependent forcing process is considered. The equation of motion of the Duffing oscillator is written as follows:

$$m\ddot{x} + c\dot{x} + k\left(x + \beta x^3\right) = f(t),$$

(3.1)

where the forcing process is assumed as a harmonic load:

$$f(t) = a \cdot \sin(\overline{\omega}t).$$

(3.2)

Uncertainties of the system are assumed to be concentrated in the stiffness k of the system; the nonlinear restoring term in (3.1) can be written as follows:

$$g(x, \dot{x}, t; k) = c\dot{x} + k\left(x + \beta x^3\right),$$

$$(3.3)$$

and, consequently, the equation of motion in general form, accordingly to (2.1), can be rewritten as follows:

$$x(t, k) = \sum_{j=1}^{N_\varphi} w_j(t)\varphi_j(k),$$

$$(3.4)$$

Based on the proposed approach (2.10) can be particularised as follows:

$$x(t, k) = \sum_{j=1}^{N_\varphi} w_j(t)\varphi_j(k),$$

$$(3.5)$$

where $\mathbf{w}(t)$ is a time-dependent vector and $\varphi j(k)$ are assumed as Gaussian-shaped RB functions depending only on the random stiffness k of the mechanical system. Considering (2.12), the solution of the original problem of (3.1) can be moved to the solution of a set of coupled deterministic nonlinear equations, and the statistics of the response process (mean value and correlation function) can be directly evaluated by means of the following:

$$\mu_x(t) = E[x(t, l)] = \sum_j w_j(t)E[\varphi_j(k)],$$

$$\sigma_x^2(t_1, t_2) = \sum_{i,j} w_i(t_1)E[(\varphi_i(k) - E[\varphi_i(k)])(\varphi_j(k) - E[\varphi_j(k)])]w_j(t_2).$$

$$(3.6)$$

To analyse the problem, the following parameters have been assumed: a=300 N; ω=15 rad/sec; Ek=800 N/cm; β=0.021/(cm^2); c=2.85 Nsec/cm; m=200 Nsec^2/cm. To check the efficiency of the approach, several tests with increasing number of RBFs ($N\varphi$) have been carried out considering 25 sec of analysis, and the results have been then compared with MCS. A set of 500 simulations (direct numerical integration of the differential equation of motion (3.1)) has been carried out by generating a vector of random values of the variable (assuming a coefficient of variation of 5%).

As a first case, 2-RBFs ($N\varphi$=2) have been considered; at each time step, a system of $N\varphi \times N$ equations has to be solved. The proposed approximation with $N\varphi$=2 matches quite well the MC simulation with respect to the mean value of displacement and velocity (Figures 1 and 2). On the contrary, if the standard deviation of both displacement and velocity are plotted together with MCS results, it is possible to see that the approximation is not able to reproduce the response process (Figures 3 and 4). The difficulty of the 2-RBFs approximation to predict the system response can be highlighted analysing the system response behaviour at several instants (Figure 5).

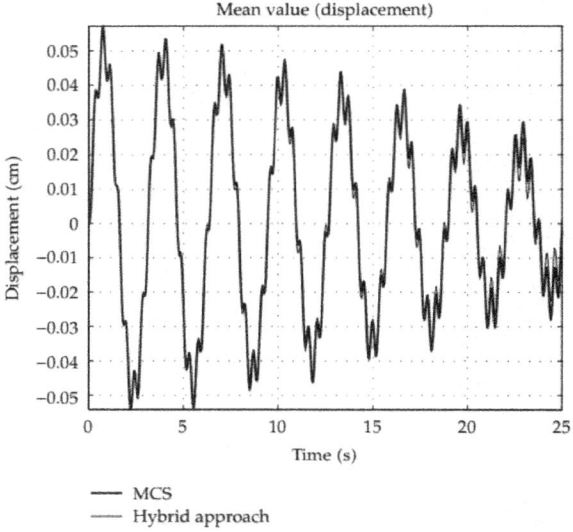

Figure 1: Mean value of displacement.

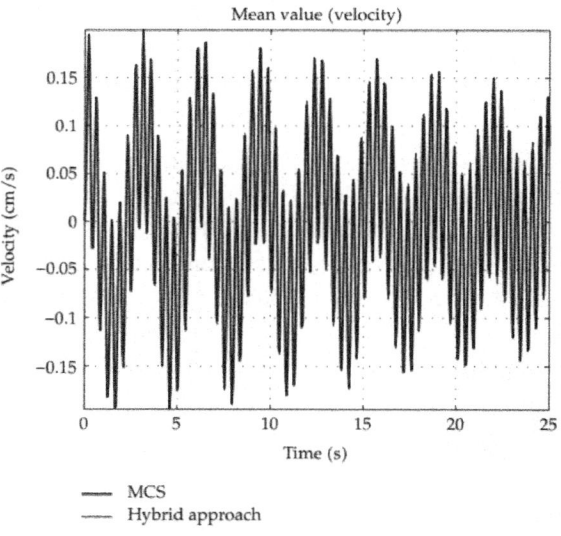

Figure 2: Mean value of velocity.

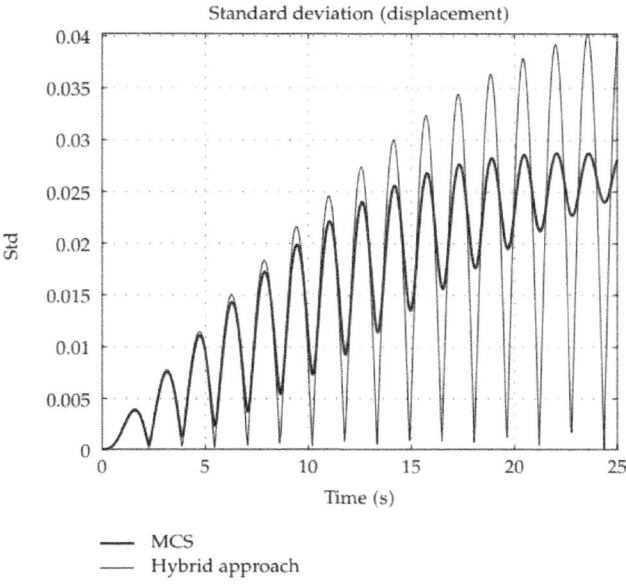

Figure 3: Standard deviation of displacement.

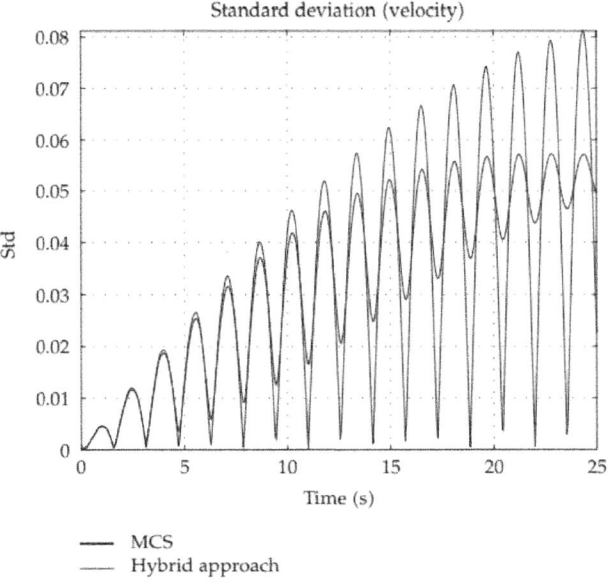

Figure 4: Standard deviation of velocity.

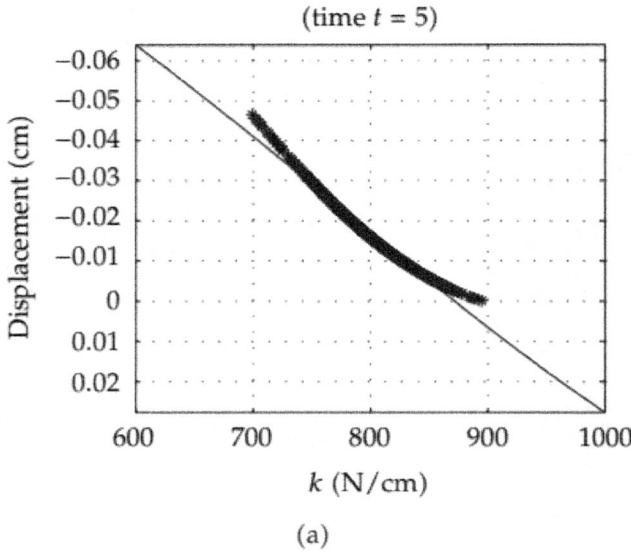

(a)

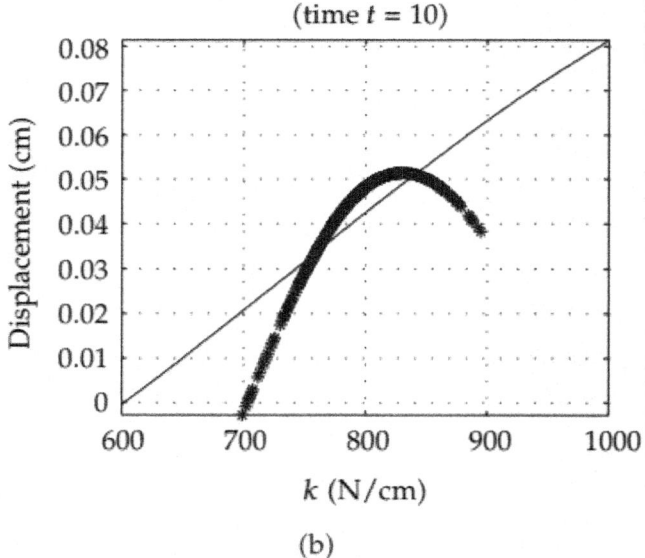

(b)

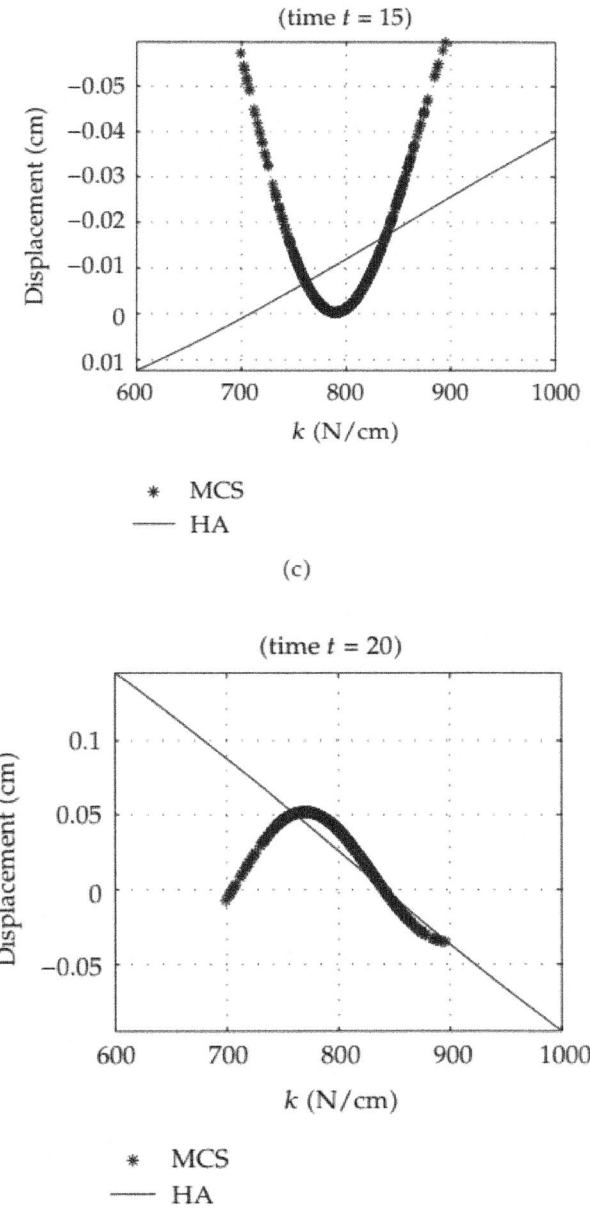

Figure 5: System response (displacement x) at different instants (5, 10, 15, and 20 sec).

Based on the above results, an approximation with 5-RBFs ($N\varphi$=5) has been considered. In figures 6, 7, 8, and9, the results of the approximation are again compared with MC simulation. In this case, the approximation function with $N\varphi$=5 is able to match the results of the simulation also with respect to the standard deviation of both displacement and velocity. The analysis of the system response behaviour at several instants (Figure 10) shows the ability of the approximation to match the actual system response.

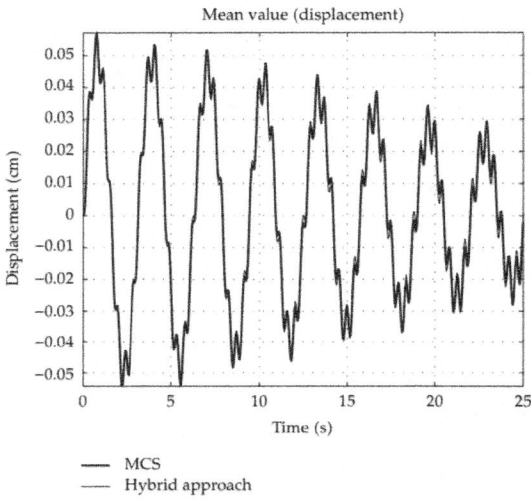

Figure 6: Mean value of displacement.

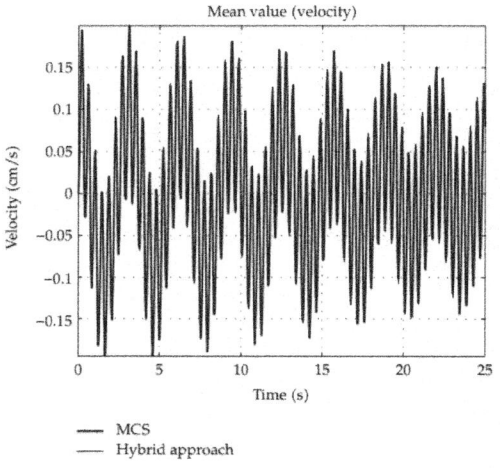

Figure 7: Mean value of velocity.

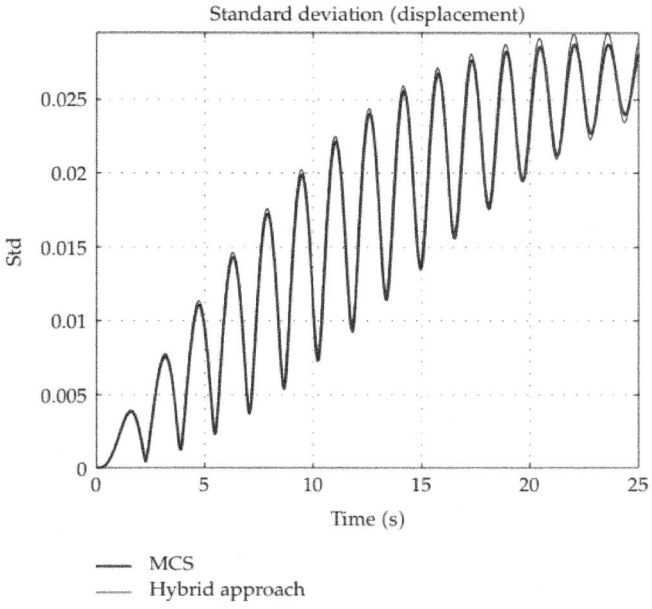

Figure 8: Standard deviation of displacement.

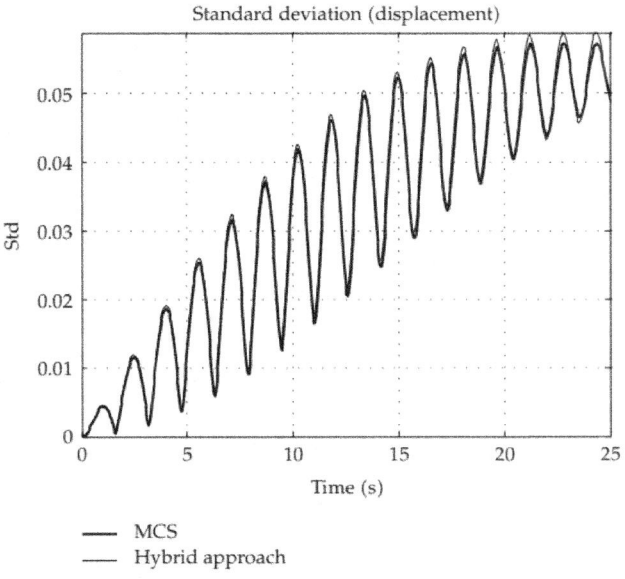

Figure 9: Standard deviation of velocity.

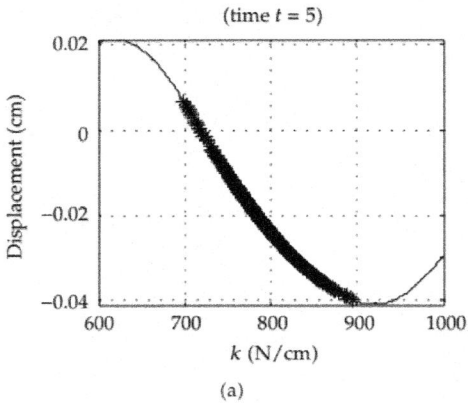

(a)

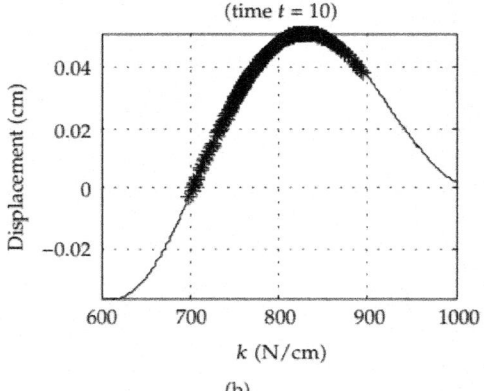

(b)

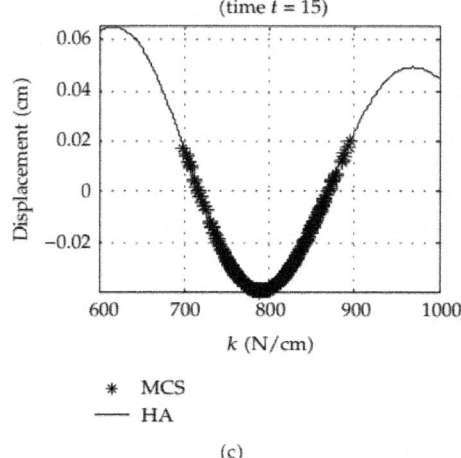

(c)

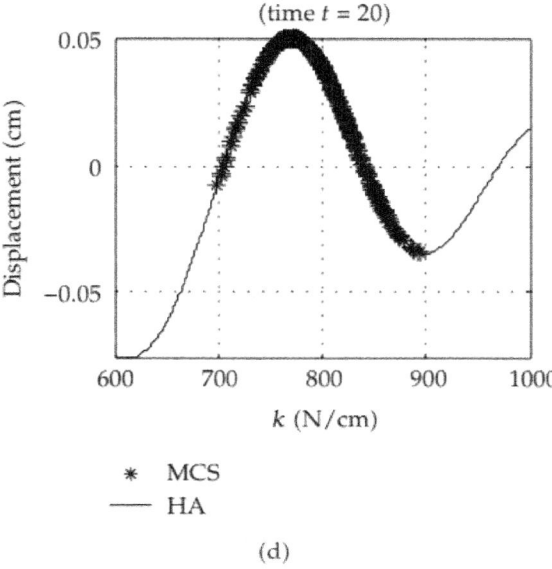

(d)

Figure 10: System response (displacement x) at different instants (5, 10, 15, and 20 sec).

Mass Pendulum

As a second case study, a mass pendulum system is considered (Figure 11) with an aim to offer an effective comparison with the existing methods to solve dynamic stochastic problems. The strongly nonlinear SDOF system is a mass pendulum system under a time-space-dependent load acting on the system mass. The parameters characterizing the system are the truss length and the mass. Naming (t) the Lagrangian parameter, the equation of motion can be written as follows:

$$ml\ddot{\theta} + mg\sin(\theta) = F(t)\cos(\theta),$$

(3.7)

where l is the pendulum length, m is the system mass, θ is the rotation, g the gravity acceleration, and $F(t)$ the external stochastic forcing load ("sin" and "cos" denote the standard trigonometric functions); dot superscripts indicate differentiation with respect to time.

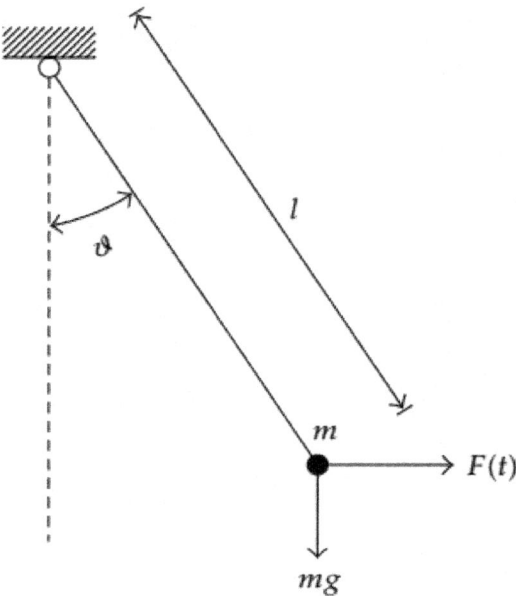

Figure 11: The mechanical nonlinear system.

It is assumed that l, the pendulum length, is an uncertain parameter; consequently, the rotation θ is depending on the length l, on the time t, and on the forcing process (t). As a first application term (t) in (3.7) has been assumed as a harmonic loading: $F(t) = F_0 \sin(\omega t)$;

$$ml\ddot{\theta} + mg\sin(\theta) = F_0 \sin(\omega t)\cos(\theta).$$

$$(3.8)$$

The stochastic dynamic problem described by (3.8) has been solved firstly by means of the perturbative approach, and results have been compared both with the proposed hybrid one (particularised to the case of deterministic load) and with the Monte Carlo simulation (MCS) assuming a set of 500 simulations.

Based on the perturbative approach, the system response is approximated by means of a series expansion with respect to the random parameters; the stochastic dynamic problem has been solved assuming following second-order Taylor series expansion of the solution about the mean values of the uncertain parameter:

$$\theta(t,l) \cong \theta(t,\mu_l) + \frac{\partial[\theta(t,\mu_l)]}{\partial l}(l-\mu_l) + \frac{1}{2}(l-\mu_l)^t\frac{\partial^2[\theta(t,\mu_l)]}{\partial l^2}(l-\mu_l).$$

$$(3.9)$$

Substituting (3.9) into (3.8) and collecting terms of the same order will yield a set of deterministic equations. In case of linear problems, this equation set has identical homogeneous parts subjected to different forcing terms, and great efficiency is achieved by the method as the system can be solved sequentially. In case of nonlinear problems, as the present case, this advantage is generally lost due to the resulting coupling of nonlinear terms; therefore, this set of equations must be solved simultaneously.

Figures 12 and 13, show the system response in terms of mean value and standard deviation obtained. Perturbative approach solutions are compared with the solution obtained by MCS. Analysing Figures 12 and 13 it is possible to observe that the solution obtained by the perturbative approach is able to match the actual system solution only up to about 20 s; afterwards, the approximation becomes unacceptable.

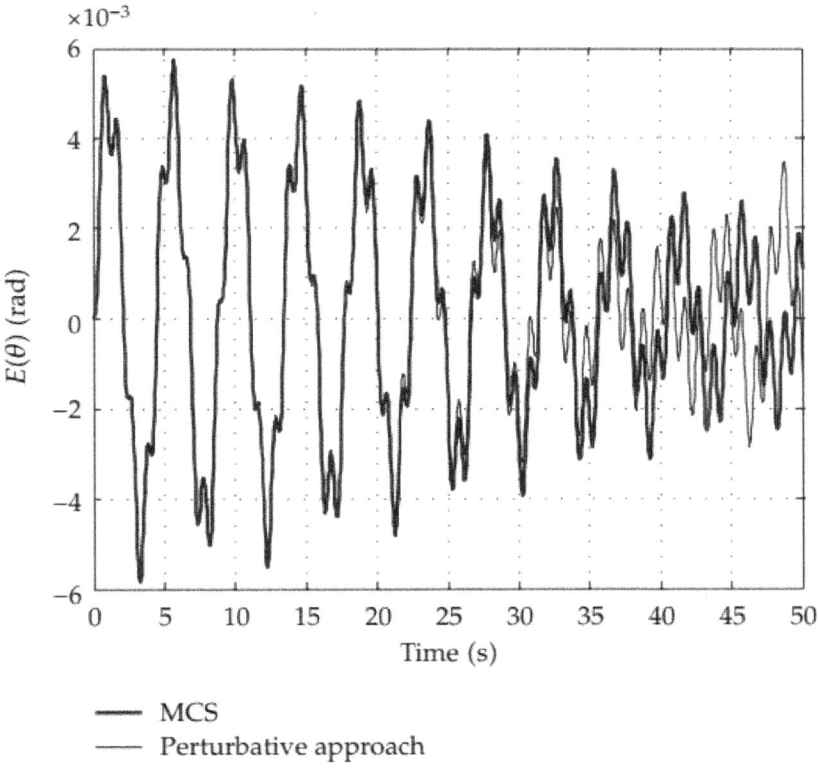

—— MCS
—— Perturbative approach

Figure 12: Mean value of θ: comparison between MCS and perturbative approach results (0–50 sec).

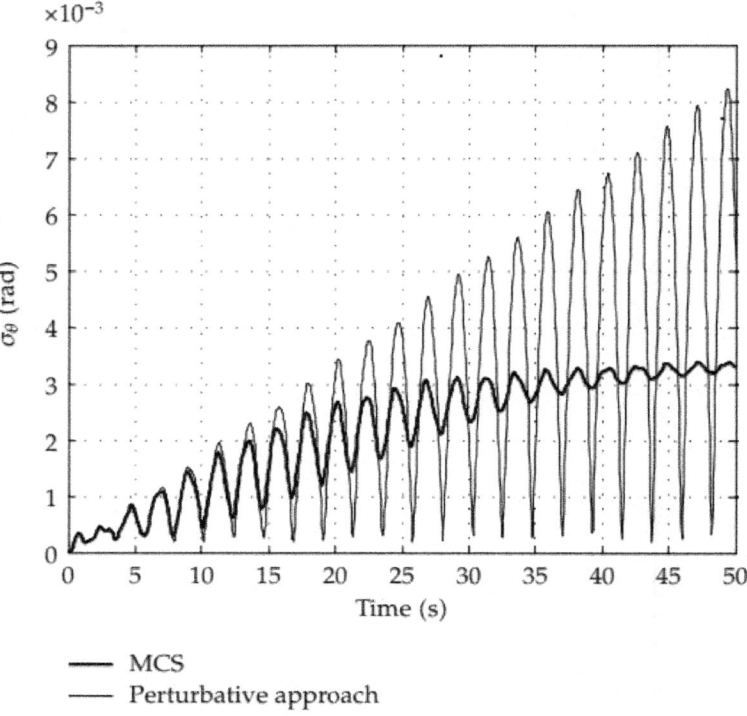

Figure 13: Standard deviation of θ: comparison between MCS and perturbative approach results (0–50 sec).

This is due to the fact that the accuracy of the results relies on the accuracy of the quadratic approximation adopted for the dependence of the response on the random parameter. Since the degree of fluctuation within this range is generally not known in advance (especially when system nonlinearity and time factors take effect) the methods can produce misleading results. In particular, the failure of the perturbative approach in the present case can be better explained analysing the dependency of the system response (rotation θ) on the pendulum length l. Figure 14 shows that the pendulum response depends approximately in a linear manner on the length of the pendulum up to about 5 s. From 10 s onwards, the quadratic approximation becomes unacceptable, and this causes the highlighted errors. It is then clear that approximation (3.9) could provide acceptable results only when the dependency of the response process by the random parameters of the system is roughly quadratic, which results into a restriction on the typology of nonlinearities that affect the system behaviour or into the variability of uncertainties.

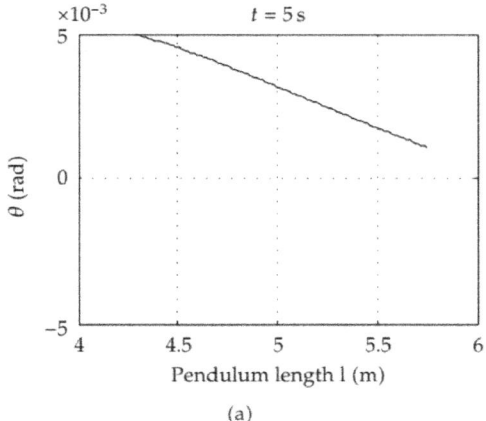

(a)

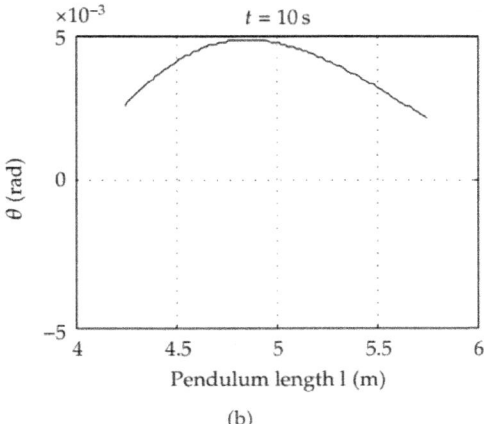

(b)

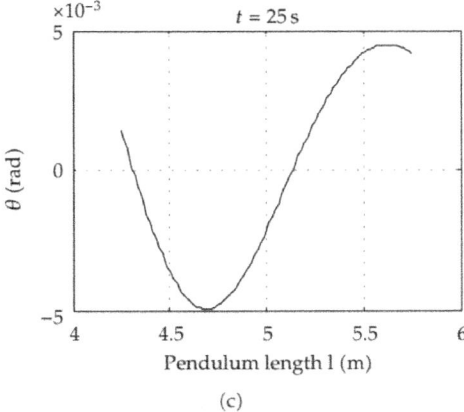

(c)

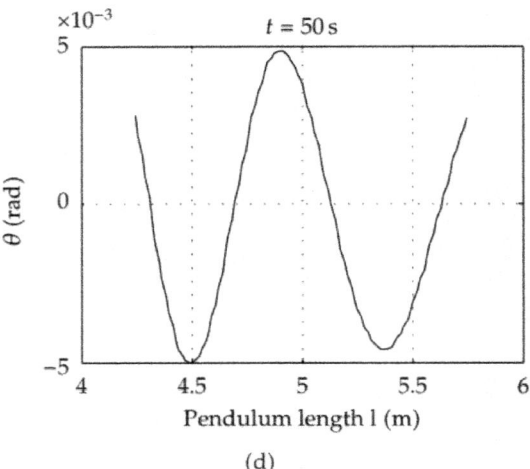

(d)

Figure 14: System response (rotation θ) at different instants (5, 10, 25, and 50 sec).

The same problem is next solved with the proposed approach (it is worthwhile noting that, in this case of deterministic loading, the hybrid approach coincides with the Galerkin/RBF); based on the assumption on deterministic loading, (2.10) is rewritten as follows:

$$\theta(t, l) = \sum_{k=1}^{N_\varphi} w_k(t)\varphi_k(l),$$

(3.10)

where $\mathbf{w}(t)$ is a time-dependent vector, $\varphi k(l)$ are functions depending only on the random parameters l of the mechanical system. Functions $\varphi(l)$ are assumed to be Gaussian-shaped RBF [21]:

$$\varphi_k(r) = \exp\left\{-\frac{r^2}{2\sigma^2}\right\}.$$

(3.11)

With these assumptions, (2.12) becomes

$$\mu_\theta(t) = E[\theta(t, l)] = \sum_k w_k(t)E[\varphi_k(l)],$$

$$\sigma_\theta^2(t_1, t_2) = \sum_{h,k} w_h(t_1)E\big[(\varphi_h(l) - E[\varphi_h(l)])(\varphi_k(l) - E[\varphi_k(l)])\big]w_k(t_2).$$

(3.12)

Figures 15 and 16 report the mean value and standard deviation of the rotation θ, respectively, comparing the solution of the proposed approach with

MCS; the proposed approximation is able to follow the system response even in the range where the perturbative approach fails. Furthermore, if the response of the pendulum is analysed at several time instants (Figure 17), the proposed method shows its capacity to approximate with confidence the actual system behaviour.

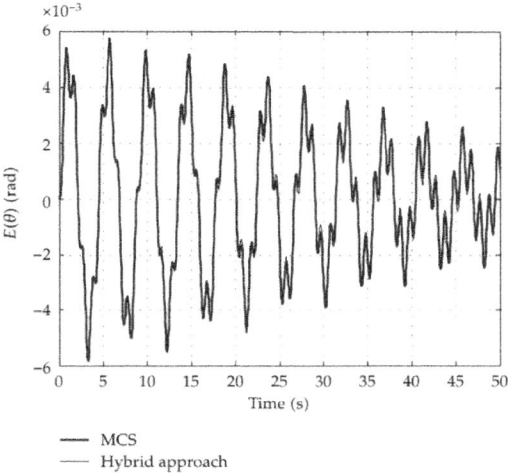

Figure 15: Mean value of θ: comparison between MCS and hybrid approach results (0–50 sec).

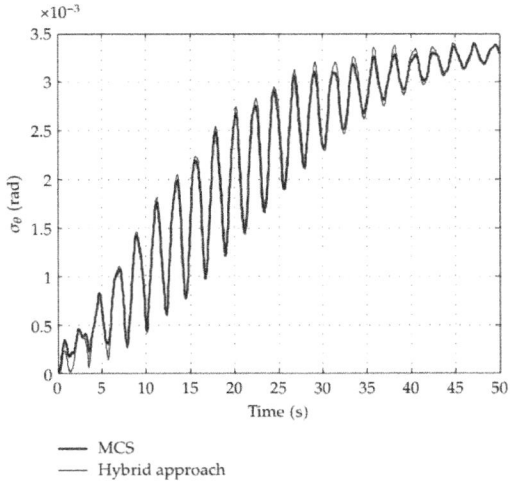

Figure 16: Standard deviation of θ: comparison between MCS and hybrid approach results (0–50 sec).

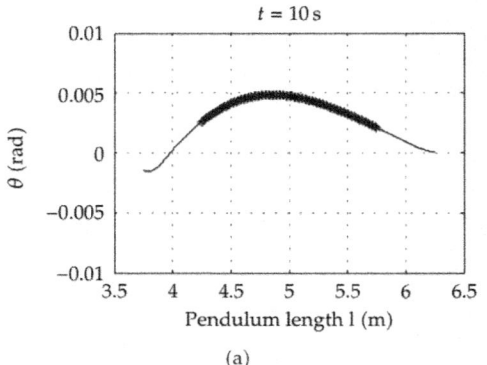

(a)

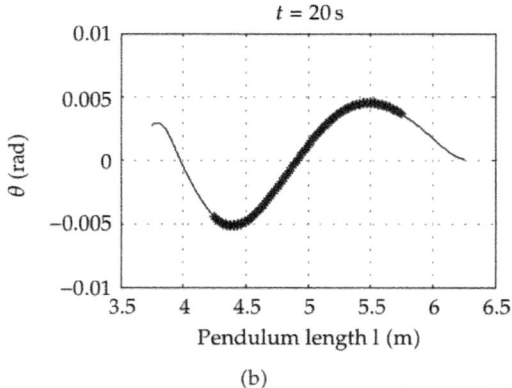

(b)

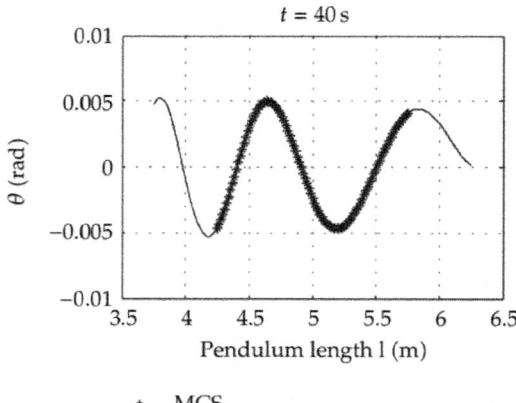

* MCS

— HA

(c)

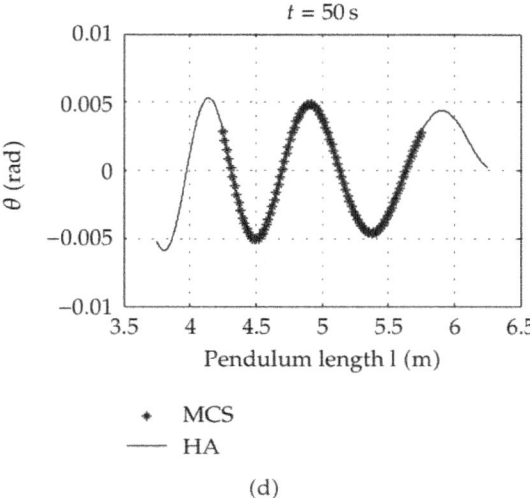

(d)

Figure 17: System response (rotation θ) at different instants (10, 20, 40, and 50 sec): comparison between MCS and hybrid approach (HA) approximation.

Stochastic Loading

Mass Pendulum

To detail the proposed approach with respect to the case of stochastic loading, the term (t) in (3.7) is assumed to be a stochastic load. Under very general conditions, the Karhunen-Loève (KL) expansion for the forcing process [15, 24] can be used:

$$F(t) = \sum_{h=1}^{N_\alpha} \alpha_h \cdot f_h(t),$$

(3.13)

where α_h are the KL coefficients of the forcing process. The perturbation method is adopted to approximate the dependency of the system response on the random vector collecting the terms of KL expansion. Taking into account (2.9), the following series expansion is used for the generalized displacement:

$$\theta(t, l, \boldsymbol{\alpha}) \cong \sum_{k=1}^{N_\varphi} w_k(t, \boldsymbol{\alpha}) \cdot \varphi_k(l),$$

(3.14)

where vector $\boldsymbol{\alpha}$ groups α_h terms and w_k functions in (3.14) are expressed by the perturbative approach:

$$w_k(t, \boldsymbol{\alpha}) = w_k(t, \boldsymbol{\mu_\alpha}) + \frac{\partial[w_k(t, \boldsymbol{\mu_\alpha})]}{\partial \alpha_i}(\alpha_i - \mu_{\alpha_i})$$

$$+ \frac{1}{2}(\alpha_i - \mu_{\alpha_i})\frac{\partial^2[w_k(t, \boldsymbol{\mu_\alpha})]}{\partial \alpha_j \partial \alpha_i}(\alpha_j - \mu_{\alpha_j}).$$

(3.15)

It is noteworthy to observe that also the possibility to use the KL expansion for the response process too would minimize the number of equations to be solved as this expansion is optimal [14]. Clearly, this is not possible since the covariance function of the system response is not known a priori.

Inserting (3.14) in (3.7), one obtains

$$ml\left[\sum_{k=1}^{N_\varphi}\ddot{w}_k(t, \boldsymbol{\alpha}) \cdot \varphi_k(l)\right] + mg\sin\left(\left[\sum_{k=1}^{N_\varphi}w_k(t, \boldsymbol{\alpha}) \cdot \varphi_k(l)\right]\right)$$

$$= F(t)\cos\left(\left[\sum_{k=1}^{N_\varphi}w_k(t, \boldsymbol{\alpha}) \cdot \varphi_k(l)\right]\right).$$

(3.16)

and multiplying (3.16)

by $\varphi_h(l)$ and rearranging, one obtains

$$\left[\sum_{k=1}^{N_\varphi}ml\ddot{w}_k(t, \boldsymbol{\alpha}) \cdot \varphi_k(l) \cdot \varphi_h(l)\right] + mg\sin\left(\left[\sum_{k=1}^{N_\varphi}w_k(t, \boldsymbol{\alpha}) \cdot \varphi_k(l)\right]\right) \cdot \varphi_h(l)$$

$$= F(t)\cos\left(\left[\sum_{k=1}^{N_\varphi}w_k(t, \boldsymbol{\alpha}) \cdot \varphi_k(l)\right]\right) \cdot \varphi_h(l).$$

(3.17)

Denoting the expected value operator by the symbol [·], evaluating the expected value of both members of (3.17), the following system of equations is obtained:

$$mE[l\varphi_k(l) \cdot \varphi_h(l)] \cdot \ddot{w}_k(t, \boldsymbol{\alpha}) + mgE[\sin(w_k(t, \boldsymbol{\alpha}) \cdot \varphi_k(l)) \cdot \varphi_h(l)]$$

$$= F(t)E[\cos(w_k(t, \boldsymbol{\alpha}) \cdot \varphi_k(l)) \cdot \varphi_h(l)],$$

(3.18)

and assuming the following notation:

$$M_{hk} = mE\left[l\varphi_k(l) \cdot \varphi_h(l)\right],$$

$$g_h = mgE\left[\sin(w_k(t, \boldsymbol{\alpha}) \cdot \varphi_k(l)) \cdot \varphi_h(l)\right],$$

$$\overline{F}_h(t, \mathbf{w}) = \left(\sum_{j=1}^{N_a} [\alpha_j \cdot f_j(t)]\right) \cdot E\left[\cos(w_k(t, \boldsymbol{\alpha}) \cdot \varphi_k(l)) \cdot \varphi_h(l)\right] = \left(\sum_{j=1}^{N_a} [\alpha_j \cdot f_j(t)]\right) \cdot T_h(\mathbf{w}),$$

$$\qquad\qquad (3.19)$$

he original problem of (3.7) can be written in vectorial form as follows:

$$\mathbf{M}\ddot{\mathbf{w}} + \mathbf{g}(\mathbf{w}) = \overline{\mathbf{F}}(t, \mathbf{w}), \qquad\qquad (3.20)$$

where the effects of the load randomness have been evaluated by means of the perturbation method, and the effects of the randomness of the uncertain parameter l are taken into account by means of the Galerkin approach. Using (3.15), evaluating (3.20) with respect to the mean value of the KL coefficients α_h of the forcing process, and collecting terms, it is possible to obtain the solving systems of equations.

The first term in the expansion is the response calculated for mean value of the KL coefficients, so that

$$\mathbf{M}\underline{\ddot{\mathbf{w}}} + \mathbf{g}(\underline{\mathbf{w}}) - \overline{\mathbf{F}}(t, \mathbf{w}) = 0,$$

$$M_{hk}\underline{\ddot{w}}_k + mgS_h - \left[\sum_{l=1}^{N_a} \mu_{\alpha_l} \cdot f_l(t)\right] \cdot C_h = 0,$$

$$\qquad\qquad (3.21)$$

With

$$S_h = E\left[\sin(w_k \cdot \varphi_k(l)) \cdot \varphi_h(l)\right],$$

$$C_h = E\left[\cos(w_k \cdot \varphi_k(l)) \cdot \varphi_h(l)\right],$$

$$\qquad\qquad (3.22)$$

where the underbar denotes evaluation of corresponding quantities with respect to the mean value of the KL coefficients [9, 15]. The first derivatives— or, in other words, the sensitivity vectors—can be derived quite easily by differentiating with respect to the generic coefficient α_i:

$$\mathbf{M}\frac{\partial \underline{\ddot{\mathbf{w}}}}{\partial \alpha_i} + \frac{\partial \mathbf{g}(\underline{\mathbf{w}})}{\partial \mathbf{w}}\frac{\partial \underline{\mathbf{w}}}{\partial \alpha_i} - \frac{\partial \overline{\mathbf{F}}(t, \mathbf{w})}{\partial \alpha_i} = 0,$$

$$M_{hk}\underline{\ddot{w}}_{k,i} + mgC_{hm}\underline{w}_{m,i} + \left[\sum_{l=1}^{N_a} \mu_{\alpha_l} \cdot f_l(t)\right] \cdot S_{hk}\underline{w}_{k,i} - f_i(t) \cdot C_h = 0,$$

$$\qquad\qquad (3.23)$$

Where

$$S_{hm} = E\big[\sin(\underline{w}_k \cdot \varphi_k(l)) \cdot \varphi_h(l) \cdot \varphi_m(l)\big],$$

$$C_{hm} = E\big[\cos(\underline{w}_k \cdot \varphi_k(l)) \cdot \varphi_h(l) \cdot \varphi_m(l)\big].$$

(3.24)

And eventually a similar relation holds for the second derivatives:

$$\mathbf{M}\frac{\partial^2 \underline{\ddot{w}}}{\partial \alpha_i \partial \alpha_j} + \frac{\partial g(\mathbf{w})}{\partial \mathbf{w}} \frac{\partial^2 \mathbf{w}}{\partial \alpha_i \partial \alpha_j} + \frac{\partial \mathbf{w}}{\partial \alpha_i}\left[\frac{\partial^2 g(\mathbf{w})}{\partial \mathbf{w}^2} \frac{\partial \mathbf{w}}{\partial \alpha_j}\right] - \frac{\partial^2 \overline{\mathbf{F}}(t,\mathbf{w})}{\partial \alpha_i \partial \alpha_j} = 0,$$

$$M_{hk}\underline{\ddot{w}}_{k,ij} + mgC_{hm}\underline{w}_{m,ij} + \left[\sum_{l=1}^{N_\alpha}\mu_{\alpha_l} \cdot f_l(t)\right] \cdot S_{hk}\underline{w}_{k,ij} - mgS_{hml}\underline{w}_{l,j}\underline{w}_{m,i}$$

$$+f_i(t)S_{hk}\underline{w}_{k,j} - f_j(t)S_{hk}\underline{w}_{k,i} + \left[\sum_{l=1}^{N_\alpha}\mu_{\alpha_l} \cdot f_l(t)\right] \cdot C_{hkk}\underline{w}_{k,i}\underline{w}_{k,j} = 0,$$

(3.25)

With

$$S_{hmn} = E\big[\sin(\underline{w}_k \cdot \varphi_k(l)) \cdot \varphi_h(l) \cdot \varphi_m(l) \cdot \varphi_n(l)\big],$$

$$C_{hmn} = E\big[\cos(\underline{w}_k \cdot \varphi_k(l)) \cdot \varphi_h(l) \cdot \varphi_m(l) \cdot \varphi_n(l)\big].$$

(3.26)

Equations (3.21), (3.23), and (3.25) constitute a system of $N_\varphi + N_\varphi \times N_\alpha + N_\varphi \times N_\alpha^2$ equations (the degree of approximation by the hybrid Galerkin/perturbation approach) that allows to evaluate the solution of the original stochastic problem described by (2.1) by solving a set of coupled deterministic nonlinear equations. This system can be solved by means of the standard techniques available for deterministic equations. After the coefficients in the hybrid Galerkin/perturbation expansion have been calculated, they allow obtaining the realization of the response process by direct evaluation of (3.14)-(3.15). Mean value and standard deviation of system response process, taking into account the property of the KL, are calculated according to (2.12) as follows:

$$\mu_x(t) = E[x(t,b)] = \sum_{k=1}^{N_\varphi} E[w_k(t,\alpha)] \cdot E[\varphi_k(b)]$$

$$= \sum_{k=1}^{N_\varphi}\left(\underline{w}_k(t,\mu_\alpha) + \frac{1}{2}\sum_{i,j}^{N_\alpha}E\big[(\alpha_i - \mu_{\alpha_i})(\alpha_j - \mu_{\alpha_j})\big]\frac{\partial^2 \underline{w}_k}{\partial \alpha_i \partial \alpha_j}\right) \cdot E[\varphi_k(b)],$$

$$\sigma_x^2 = \sum_{h=1}^{N_\varphi} E\big[w_h^2\big] \cdot E\big[\varphi_h^2\big] - E[w_h]^2 \cdot E[\varphi_h]^2.$$

(3.27)

Based on the results discussed in previous paragraphs, the case of stochastic loading is analysed. Preliminarily a set of 500 simulations has been carried out by generating a vector of random values of the variable l with assigned mean and variance. A Gaussian distribution has been assumed, considering a mean value $\mu l = 5$ m and a coefficient of variation for the uncertain parameter of 5% and 10%. The response of the system has been obtained by direct numerical integration of the differential (3.7) governing the problem. The equations of motions were integrated in the case of a wind-pressure-type random excitation, with a Ficthl-Mc Vehil [25] spectral density function defined by

$$\frac{f S_{FF}(f)}{\sigma_F^2} = \frac{A f}{\left(1 + B f^{\beta}\right)^{\alpha}},$$

$$B = 0.0216, \quad \alpha = 1.3608, \quad \beta = 1.7034,$$

$$A = \frac{B^{1/\beta} \Gamma(\alpha)}{\Gamma(\alpha - 1/\beta) \Gamma(1 + 1/\beta)} = 0.0879.$$

$$(3.28)$$

The excitation was described by 200 Karhunen-Loève eigenfunctions (Figure 18 reports a typical realization). Integration lasted for 50 seconds.

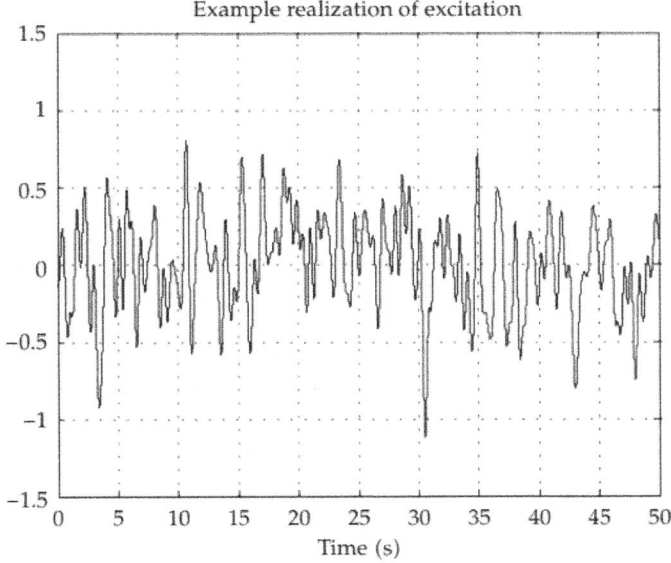

Figure 18: A realization of the excitation process employing 200 terms in Karhunen-Loève expansion.

In order to estimate the quality of the approximation offered by the proposed hybrid Galerkin/perturbation approach, several cases are investigated where different values of the radial basis functions numbers and KL terms in (3.14) and (3.15) have been taken into account. As for the simulation task, a time-length of 50 seconds of the analysis has been considered. The outputs of the outlined procedure, the sensitivity coefficients, are obtained by integration of (3.21)–(3.23)–(3.25). Indicating with N_{φ} the number of RBFs adopted and with $N\alpha$ the number of KL terms, at each time steps a system of $N_{\varphi} + N_{\varphi} \times N_{\alpha} + N_{\varphi} \times N_{\alpha}^2$ equations has to be solved. After integration, upper and lower bounds were obtained for both rotation and angular velocity of the pendulum by means of [9]:

Upperbound: $\sup(\theta) \approx \mu_{\theta} + 3\sigma_{\theta}$,

Lowerbound: $\inf(\theta) \approx \mu_{\theta} - 3\sigma_{\theta}$.

$$(3.29)$$

As a first test, a case where wind load is approximated by 200 terms in KL summation (3.13) assuming a coefficient of variation of 5%, has been considered. Several tests with different number of RBFs have been taken into account.

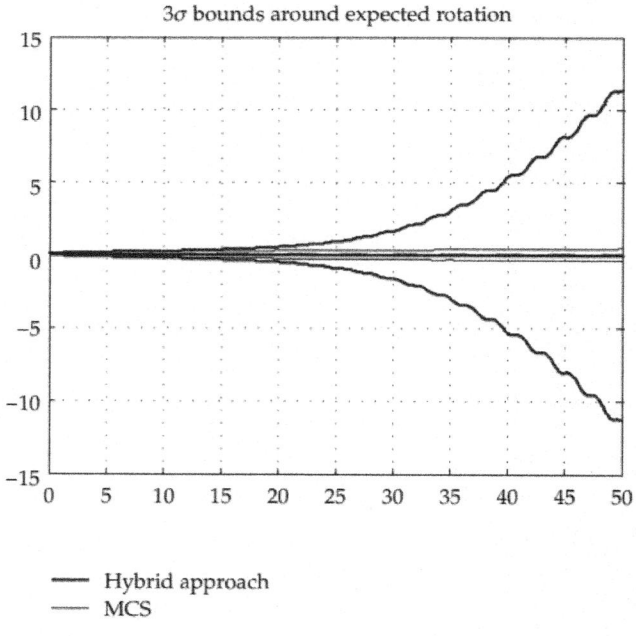

Figure 19: Upper and lower bounds for pendulum rotation ($N\varphi=5$; $N\alpha=200$).

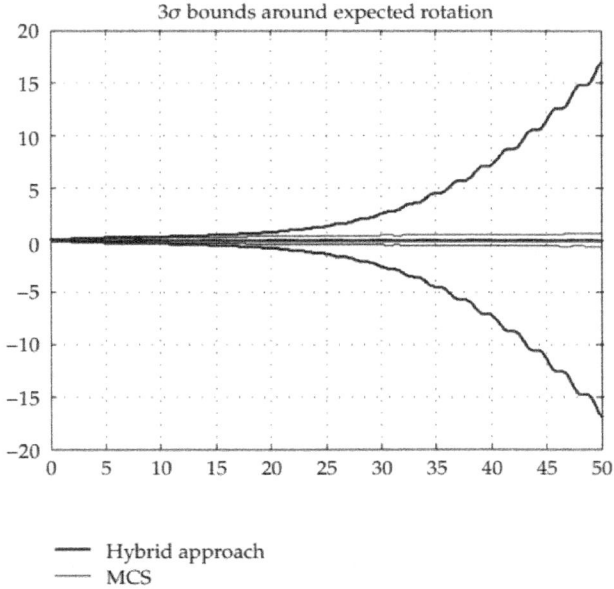

Figure 20: Upper and lower bounds for angular velocity ($N\varphi=5$; $N\alpha=200$).

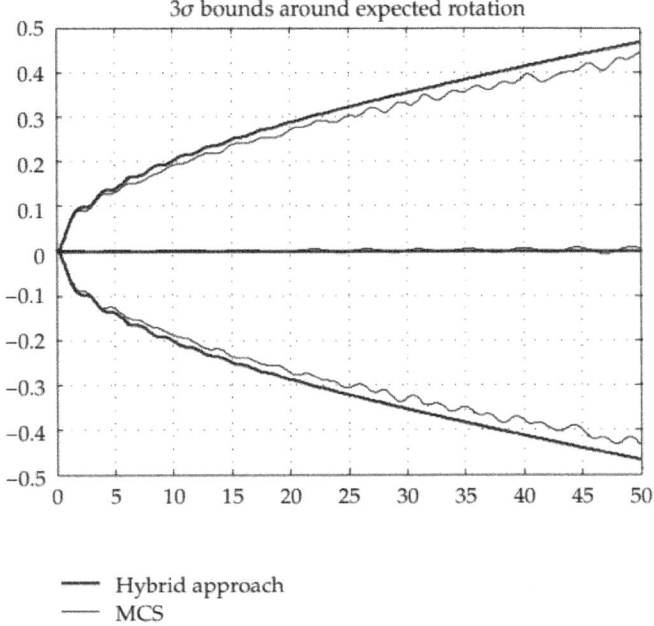

Figure 21: Upper and lower bounds for pendulum rotation ($N\varphi=20$; $N\alpha=200$).

Figures 19 and 20 report, respectively, upper and lower bounds for pendulum rotation and for angular velocity assuming 5 RBF. Results of the proposed approach compared with Monte Carlo simulation show that the level of approximation by radial basis functions is not satisfactory. Increasing the number $N\varphi$ of radial basis functions up to 10, no relevant differences appear with respect to the case with 5 RBFs. A better approximation of the PDF is, of course, obtained, but the approximation is able to reproduce the simulation only for a limited time range. A good estimation of the system response both in terms of rotation and angular velocity is possible to obtain taking into account 20 RBFs (Figures 21 and 22).

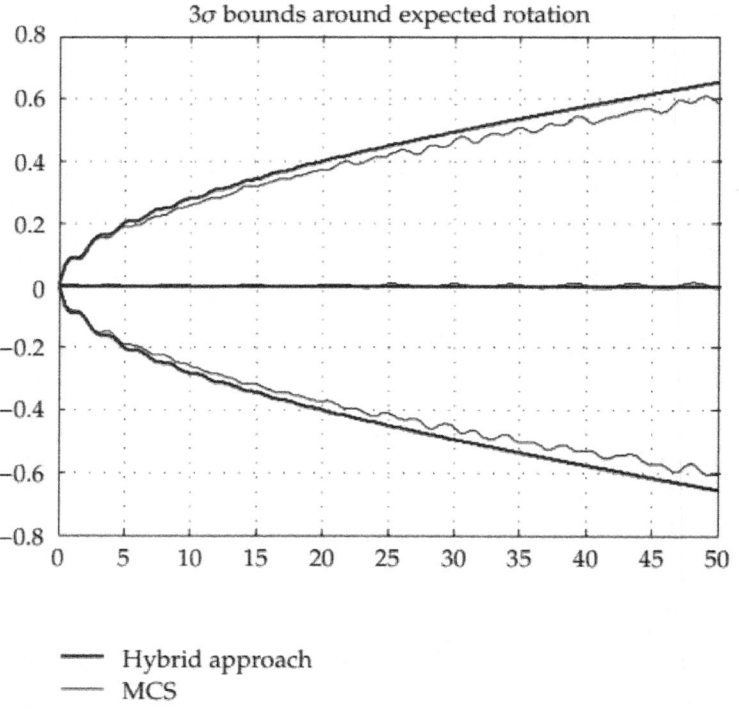

Figure 22: Upper and lower bounds for angular velocity ($N\varphi$=20; $N\alpha$=200).

As a second case, the system response with a coefficient of variation of 10% has been analysed (the numbers of N_α terms have been considered again 200). Investigations have been made with respect to the number N_φ of radial basis functions that varies from 5 to 40. Taking into account a small number of RBFs (N_φ=5), the results of the proposed approach are not satisfactory as shown in Figures 23 and 24 where upper and lower bounds for pendulum rotation and for angular velocity are reported compared with MCS results. Increasing the number of RBF up to 10, no significant improvement of the system response is obtained (Figures 25 and 26). It is possible to obtain a good estimation of the system response both in terms of rotation and angular velocity if the number of RBF is increased to 40. Figure 27 reports the upper and lower bounds for pendulum rotation, and Figure 28 reports the upper and lower bounds for angular velocity for this case.

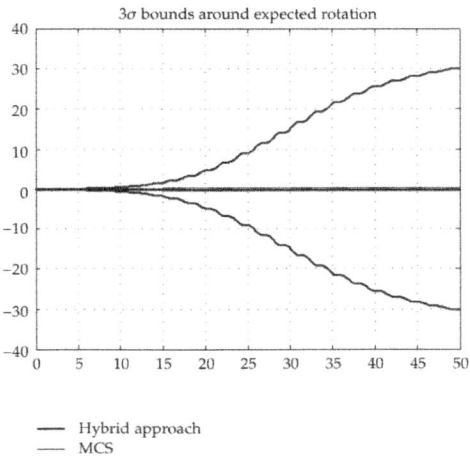

Figure 23: Upper and lower bounds for pendulum rotation (N_φ=5; N_α=200).

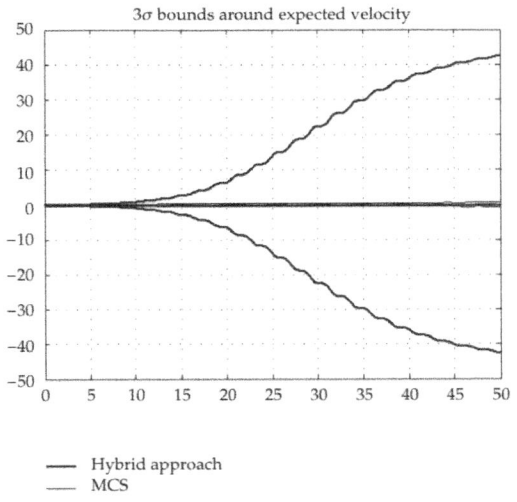

Figure 24: Upper and lower bounds for angular velocity (N_φ=5; N_α=200).

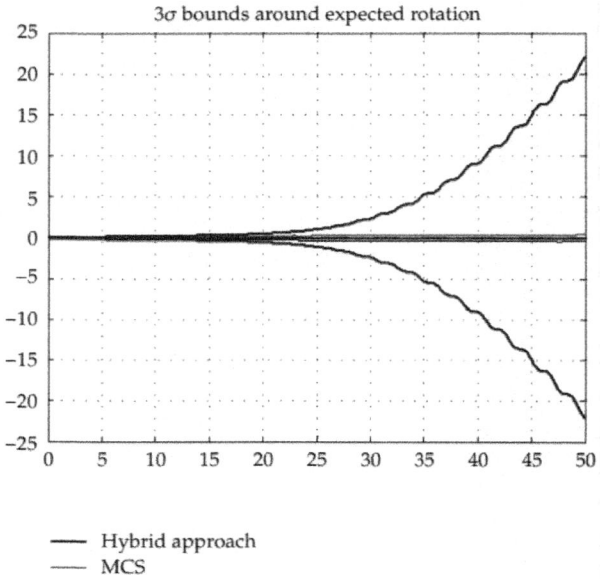

Figure 25: Upper and lower bounds for pendulum rotation (N_φ=10; N_α=200).

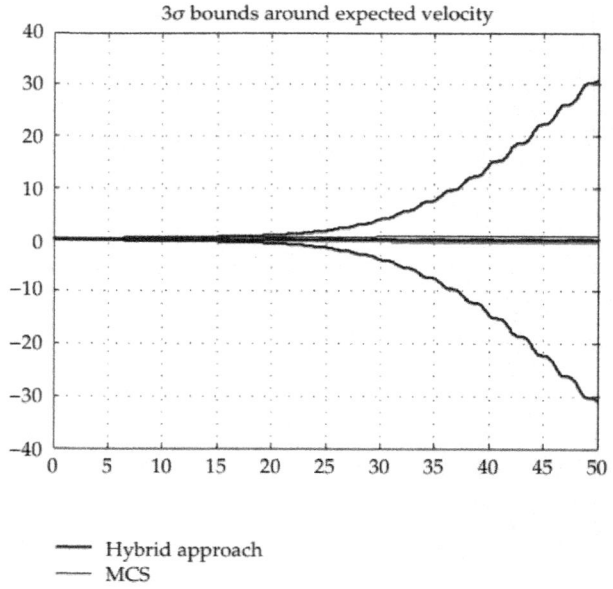

Figure 26: Upper and lower bounds for angular velocity (N_φ=10; N_α=200).

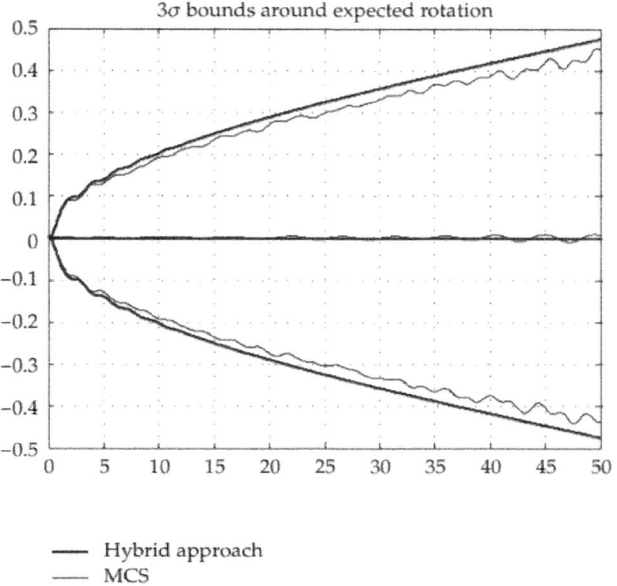

Figure 27: Upper and lower bounds for pendulum rotation (N_φ=40; N_α=200).

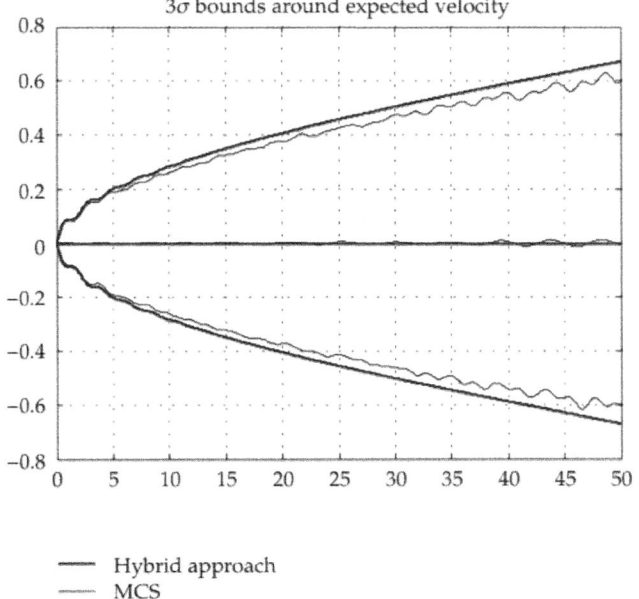

Figure 28: Upper and lower bounds for angular velocity (N_φ=40; N_α=200).

As expected, as the coefficient of variation of uncertain parameters increases, an increased number of RBFs need to be taken into account in the response approximation to match the actual system behaviour, hence increasing the overall computational cost of the proposed procedure. Nevertheless, the hybrid approach seems to be able to reproduce with sufficient care main statistics of the system response. Further developments of the current version of the proposed hybrid approach will be aimed at a proper definition of an error function to address the number of terms to be considered in the RBF approximation. It seems, in fact, that the approach is able to match the response process of arbitrarily complex nonlinear systems provided that a sufficient number of terms is included in the approximation. This is, in fact, a general property of neural networks.

CONCLUDING REMARKS

As randomness of the response process of a mechanical system can be due either to randomness in structural parameters and/or in the forcing process (f.i. wind or earthquake), the paper presents a combined hybrid Galerkin/perturbation approach based on radial basis functions for the dynamic analysis of nonlinear mechanical systems affected by randomness both in system parameters and loads. The paper suggests that the two aspects could be separated, treating the effects of randomness of mechanical parameters with a Galerkin-radial basis functions approach while the effects of the load randomness are investigated by means of a perturbation technique. This is due to the characteristics of the two methods; on the one hand, the Galerkin—RBF approach enables to evaluate with very good approximation the dependence of the structural response on the random parameters, but it increases dramatically the degrees of freedom of the examined problem. On the other hand, the perturbation method enables to evaluate a second-order expansion of the dependence of the structural response on the random parameters, thus leading to big errors in the case of strong deviations from the parabolic law. The resulting approach is quite general, and the proposed expansion for the solution contains all the necessary probabilistic information that allows to characterize the response process. As a test, the proposed procedure has been applied to a strongly nonlinear single degree of freedom problem, and the results have been compared with Monte Carlo simulation. The comparisons demonstrate that the proposed method, although limited for the time being, is a promising candidate to approach uncertain dynamic problems under stochastic loading even if it deserves further investigation to assess its efficiency with respect to other applications.

REFERENCES

1. W. D. Iwan and C. T. Huang, "On the dynamic response of non-linear systems with parameter uncertainties," International Journal of Non-Linear Mechanics, vol. 31, no. 5, pp. 631–645, 1996.

2. G. Muscolino, G. Ricciardi, and N. Impollonia, "Improved dynamic analysis of structures with mechanical uncertainties under deterministic input," Probabilistic Engineering Mechanics, vol. 15, no. 2, pp. 199–212, 2000.

3. H. Jensen and W. D. Iwan, "Response of systems with uncertain parameters to stochastic excitation,"Journal of Engineering Mechanics, vol. 118, no. 5, pp. 1012–1025, 1992.

4. M. Betti, P. Biagini, and L. Facchini, "Stochastic dynamics of complex uncertain systems by means of a conditional—Galerkin—RBF approach," in Proceedings of the 6th European Conference on Structural Dynamics (EURODYN ‹05), pp. 847–852, Paris, France, September 2005.

5. R. G. Ghanem and P. D. Spanos, Stochastic Finite Elements: A Spectral Approach, Springer, New York, NY, USA, 1991.

6. M. Di Paola, "Probabilistic analysis of truss structures with uncertain parameters (virtual distortion method approach)," Probabilistic Engineering Mechanics, vol. 19, no. 4, pp. 321–329, 2004.

7. V. Papadopoulos and M. Papadrakakis, "The effect of material and thickness variability on the buckling load of shells with random initial imperfections," Computer Methods in Applied Mechanics and Engineering, vol. 194, no. 12-16, pp. 1405–1426, 2005.

8. N. D. Lagaros and V. Papadopoulos, "Optimum design of shell structures with random geometric, material and thickness imperfections," International Journal of Solids and Structures, vol. 43, no. 22-23, pp. 6948–6964, 2006.

9. W. K. Liu, T. Belytschko, and A. Mani, "Probabilistic finite elements for nonlinear structural dynamics,"Computer Methods in Applied Mechanics and Engineering, vol. 56, no. 1, pp. 61–81, 1986.

10. F. Yamazaki, M. Shinozuka, and G. Dasgupta, "Neumann expansion for stochastic finite element analysis," Journal of Engineering Mechanics, vol. 114, no. 8, pp. 1335–1354, 1988.

11. P. D. Spanos and R. Ghanem, "Stochastic finite element expansion for random media," Journal of Engineering Mechanics, vol. 115, no. 5, pp. 1035–1053, 1989.

12. I. Elishakoff, Y. J. Ren, and M. Shinozuka, "Improved finite element method for stochastic problems,"Chaos, Solitons and Fractals, vol. 5, no. 5, pp. 833–846, 1995.

13. G. Muscolino, G. Ricciardi, and N. Impollonia, "Improved dynamic analysis of structures with mechanical uncertainties under deterministic input," Probabilistic Engineering Mechanics, vol. 15, no. 2, pp. 199–212, 2000.

14. R. Ghanem, "Hybrid stochastic finite elements: coupling of spectral expansions with monte carlo simulations," Journal of Applied Mechanics (ASME), vol. 65, no. 4, pp. 1004–1009, 1998.

15. S. Chiostrini and L. Facchini, "Response analysis under stochastic loading in presence of structural uncertainties," International Journal for Numerical Methods in Engineering, vol. 46, no. 6, pp. 853–870, 1999.

16. D. S. Broomhead and D. Lowe, "Multivariable functional interpolation and adaptive networks," Complex Systems, vol. 2, pp. 269–303, 1988.

17. T. Poggio and F. Girosi, "Networks for approximation and learning," Proceedings of the IEEE, vol. 78, no. 9, pp. 1481–1497, 1990.

18. L. Facchini, "Analysis of disordered structures by means of radial basis function networks," in Proceedings of the 9th International Conference on Structural Safety and Reliability (ICOSSAR ‹05), pp. 2305–2310, Rome, Italy, June 2005.

19. L. Facchini and S. Chiostrini, "La stima della risposta di un sistema meccanico a parametri incerti mediante funzioni RBF," in Proceedings of the 2nd Workshop Problemi di Vibrazioni nelle Strutture Civili e nelle Costruzioni Meccaniche, pp. 187–196, Perugia, Italy, June 2004.

20. S. Haykin, Adaptive Filter Theory, Prentice Hall, Upper Saddle River, NJ, USA, 1996.

21. B. Gotovac and V. Kozulic, "On a selection of basis functions in numerical analyses of engineering problems," International Journal for Engineering Modelling, vol. 12, no. 1–4, pp. 25–41, 1999.

22. N. Mai-Duy and T. Tran-Cong, "Numerical solution of differential equations using multiquadric radial basis function networks," Neural Networks, vol. 14, no. 2, pp. 185–199, 2001.

23. F. Facchini, M. Betti, and P. Biagini, "Random dynamics of simple disordered masonry structures," inProceedings of the Random dynamics of simple disordered masonry structures, pp. 263–271, Rhodes, Greece, June 2006.

24. R. G. Ghanem and P. D. Spanos, "Spectral stochastic finite-element formulation for reliability analysis,"Journal of Engineering Mechanics, vol. 117, no. 10, pp. 2351–2372, 1991.

25. L. Facchini, "The numerical simulation of Gaussian cross-correlated wind velocity fluctuations by means of a hybrid model," Journal of Wind Engineering and Industrial Aerodynamics, vol. 64, no. 2-3, pp. 187–202, 1996.

Chapter 10

STOCHASTIC DYNAMICS OF A DRILL-STRING WITH UNCERTAIN WEIGHT-ON-HOOK

T. G. Ritto[1];C. Soize[2];and Rubens Sampaio[3]

[1]PUC-Rio Department of Mechanical Engineering 22453-900 Rio de Janeiro, RJ, Brazil. Université Paris-Est Lab. de Modélisation et Simulation Multi-Echelle 77454 Marne-la-Vallée, France

[2]Université Paris-Est Lab. de Modélisation et Simulation Multi-Echelle 77454 Marne-la-Vallée, France

[3]PUC-Rio Department of Mechanical Engineering 22453-900 Rio de Janeiro, RJ, Brazil

ABSTRACT

A drill-string is a slender structure that turns and drills into the rock in search of oil. There are many sources of uncertainties in this complex dynamical system. However, this article is concerned only with uncertainties in the weight-on-hook, which is the supporting force exerted by the hook at the top. A probabilistic model is constructed for the random variable related to the weight-on-hook using the Maximum Entropy Principle, and the random response of the system is computed through Monte Carlo simulations. The idea is to understand how the performance of the system (which is measured by the rate of penetration) is affected by the uncertainties of the weight-on-hook. The continuous system analyzed is discretized by means of the Finite Element Method and a computer code is developed to do the simulations.

INTRODUCTION

In a drilling operation there are many sources of uncertainties, such as the material properties of the column and of the drilling fluid; the dimensions of the system, especially of the borehole; the fluid-structure interaction; and the bit-rock interaction. To have an improved computational model with better predictability capacity, uncertainties should be quantified and included in the computational model, which then becomes a stochastic computational model (if the probability theory is used to quantify the uncertainties).

This paper is concerned with uncertainties in the weight-on-hook because it is one of the three parameters that are continuously controlled in a drilling operation. (The other two parameters are the rotational speed at the top and the inlet fluid velocity). Figure 1 shows the general scheme of the system analyzed. The motor torque is taken into account as a constant rotational speed at the top (Ω_x) and the weight-on-hook (which is the supporting force, w) is constant. The bit reacts to the dynamics of the column with the torque t_{bit} and the force f_{bit}.

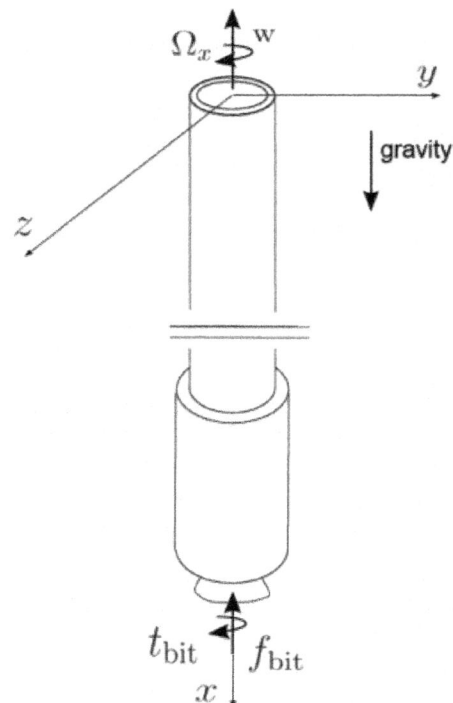

Figure 1: General scheme of the system analyzed.

The probability theory, which is a powerful tool to model uncertainties, is used in the present analysis. In a general way, there are two types of uncertainties: (1) one type related to the randomness of some parameter (aleatory uncertainties) and (2) other type related to the lack of knowledge of a given phenomenon (epistemic uncertainties), which is related to modeling errors. One way to take into account epistemic uncertainties is to use the nonparametric probabilistic approach (Soize, 2000) because it is able to model both parameter uncertainties as well as model uncertainties.

The analysis of the dynamics of flexible structures considering the deterministic and the stochastic problem (Sampaio and Ritto, 2008) is having an increasing attention of the scientific community over the years. Ritto et al. (2008), for example, analyze parameter and model uncertainties in the boundary condition of a beam.

In Tucker and Wang (1999), Khulief and AL-Naser (2005), Sampaio et al. (2007), Piovan and Sampaio (2009), for instance, the deterministic model of a drill-string system is analyzed. There are few articles treating the stochastic problem of the drill-string dynamics; see, for instance, Spanos et al. (1997, 2009), Ritto et al. (2009, 2010). In Kotsonis and Spanos (1997) a random weight-on-bit is analyzed in a simple two-degrees-of-freedom drillstring model, and in Spanos et al. (2009) lateral forces at the bit are modeled as stochastic. In Ritto et al. (2009) a probabilistic model (nonparametric probabilistic approach) is proposed for the bit-rock-interaction model, and in Ritto et al. (2010) a robust optimization problem is performed, where model uncertainties are modeled using the nonparametric probabilistic approach.

In the present paper, the focus of the stochastic analysis is only on the uncertain weight-on-hook, although there are other sources of uncertainties, such as the ones mentioned above. It is important to analyze a complex problem incrementally, in a way that the final result is not obscured by the influence of many factors at the same time. The parametric probabilistic approach is used to model the uncertainties in the weight-on-hook, and the probability density function of the random variable related to the weight-on-hook is constructed by the means of the Maximum Entropy Principle (Shannon, 1948; Jaynes, 1957a,b).

The strategy goes the following way. After the construction of the probability density function of the random weight-on-hook, a random generator is used to generate a value for the weight-on-hook for each Monte Carlo simulation (Rubinstein, 2007). The number of Monte Carlo simulations is chosen analyzing a convergence curve. The random response can finally be analyzed by the computations of some statistics, such as mean and confidence intervals.

In the first Section of the paper, the discretized system is presented and the element matrices are depicted, then, in the second Section, the reduced-order model is developed. The probabilistic model for the weight-on-hook is constructed in the third Section and the stochastic dynamical system is presented in the fourth Section. The numerical results are presented in the fifth Section and, finally, the concluding remarks are presented in the last Section.

Nomenclature

A	=	*cross sectional area, m^2*
conv	=	*convergence function, m^2*
D	=	*diameter, m*
E	=	*Young modulus, Pa*
g	=	*acceleration of gravity, m/s^2*
G	=	*shear modulus, Pa*
I	=	*moment of inertia of the transversal section, m^4*
S	=	*Shannon entropy measure*
t	=	*time, s*
u	=	*displacement in x-direction, m*
Z	=	*regularizing function*
f	=	*force vector, N, N.m*
N	=	*shape functions, m*
u	=	*displacement vector, m, rad*
$[C]$	=	*damping matrix, N.s/m, N.s.m*
$[K]$	=	*stiffnes matriz, N/M, N.m*
$[M]$	=	*mass matrix, kg, kg.m^2*

Greek Symbols

$1_B(x)$ = *assumes value 1 if x belongs to B and 0 otherwise*

δ	=	*coeffictient of variation*
ρ	=	*density, kg/m^3*
Ω_x	=	*rotational speed at x = 0, rad/s*
σ	=	*standard deviation, N*
θ_x	=	*rotational about x-axis, rad/s*
$[\Phi]$	=	*modal basis, m, rad*

Subscripts

br	bit-rock
f	fluid
g	geometric (for $[K]$) and gravity (for **f**)
i	inside
o	outside
p	polar
S	static response
u	displacement in x-direction
θ_x	rotational about x-axis

FINAL DISCRETIZED SYSTEM

The present work is concerned with the stochastic response of the system when the weight-on-hook (axial force) is random. To focus the attention of the analysis on the stochastic problem, and to speed up the numerical simulations, a simplification of the model found in Ritto et al. (2009) is used for the present analysis. The lateral displacements are assumed to be small and, therefore, they are neglected; and the system is linearized about the prestressed state $\mathbf{u}_s$. This state is computed through the equation, $\mathbf{u}_s = [K]^{-1} (\mathbf{f}_g + \mathbf{f}_c + \mathbf{f}_f)$ where [K] is the stiffness matrix of the system, $\mathbf{f}_g$ is the gravity force, $\mathbf{f}_c$ is the concentrated reaction force at the bit, and $\mathbf{f}_f$ is the fluid axial force. The final discretized system considering the prestressed state is written as: in which the response $\bar{\mathbf{u}}(= \bar{\mathbf{u}} - \bar{\mathbf{u}}_s)$ is represented in a subspace $V_m \subset \mathbb{R}^m$, where m equals the number of degrees of freedom of the system. [M], [C], and [K] are the classical mass, damping and stiffness matrices, $[K_g (\mathbf{u}_s)]$ is the geometric stiffness matrix (due to the finite strain formulation), $\mathbf{g}$ is the force due to the Dirichlet boundary condition (imposed rotational speed at the top) and $\mathbf{f}_{br}$ are the forces due to the bit-rock interactions.

$$[M]\ddot{\bar{\mathbf{u}}}(t) + [C]\dot{\bar{\mathbf{u}}}(t) + \left([K] + [K_g(\mathbf{u}_s)]\right)\bar{\mathbf{u}}(t) =$$
$$= \mathbf{g}(t) + \mathbf{f}_{br}\left(\dot{\bar{\mathbf{u}}}(t)\right) \tag{1}$$

The proportional damping matrix is constructed a posteriori [C] = [C] = a[M]+b([K]+[K_g (\mathbf{u}_s)]), where a and b are positive constants. The finite element approximation of the displacement fields is written as

$$u(\xi,t) = \mathbf{N}_u(\xi)\mathbf{u}_e(t)$$
$$\theta_x(\xi,t) = \mathbf{N}_{\theta_x}(\xi)\mathbf{u}_e(t) \tag{2}$$

where u is the axial displacement, θ_x is the rotation about the x-axis, $\xi = x/l_e$ is the element coordinate, l_e is the element length, $\mathbf{N}$ are the shape functions

$$\mathbf{N}_u = \begin{bmatrix} (1-\xi) & 0 & \xi & 0 \end{bmatrix},$$
$$\mathbf{N}_{\theta_x} = \begin{bmatrix} 0 & (1-\xi) & 0 & \xi \end{bmatrix}, \tag{3}$$

and the element node displacements are

$$\mathbf{u}_e = \begin{bmatrix} u_1 & \theta_{x1} & u_2 & \theta_{x2} \end{bmatrix}^T, \tag{4}$$

where the exponent T means transposition.

The element forces and matrices are depicted in the sequence. The mass $[M]^{(e)}$ and stiffness $[K]^{(e)}$ element matrices are written as

$$[M]^{(e)} = \int_0^1 \left[\rho A \left(\mathbf{N}_u^T \mathbf{N}_u \right) + \rho I_p \left(\mathbf{N}_{\theta_x}^T \mathbf{N}_{\theta_x} \right) \right] l_e d\xi,$$

(5)

and

$$[K]^{(e)} = \int_0^1 \left[\frac{EA}{l_e} \left(\mathbf{N}_u'^T \mathbf{N}_u' \right) + \frac{GI_p}{l_e} \left(\mathbf{N}_{\theta_x}'^T \mathbf{N}_{\theta_x}' \right) \right] d\xi,$$

(6)

where ρ is the density, A is the cross sectional area, I_p is the polar moment of inertia, E is the elasticity modulus, G is the shear modulus, and the space derivative $(d/d\,\xi)$ is denoted by $(\cdot)$. The element geometric stiffness matrix is written as

$$\begin{aligned}
\left[K_g(u) \right]^{(e)} &= \int_0^1 \Big[\left(\mathbf{N}_u'^T \mathbf{N}_u' \right) \left(3EAu' + 1.5EAu'^2 + \right. \\
&+ 0.5EI_p \theta_x'^2 \Big) + \left(\mathbf{N}_u'^T \mathbf{N}_{\theta_x}' \right) \left(EI_p \theta_x' + EI_p \theta_x' u' \right) + \\
&+ \left(\mathbf{N}_{\theta_x}'^T \mathbf{N}_u' \right) \left(EI_p \theta_x' + EI_p \theta_x' u' \right) + \left(\mathbf{N}_{\theta_x}'^T \mathbf{N}_{\theta_x}' \right) \left(EI_p u' + \right. \\
&+ 0.5EI_p u'^2 + 1.5EI_{p4} \theta_x'^2 + 3I_{22} \theta_x'^2 \Big) \Big] \frac{1}{l_e} d\xi,
\end{aligned}$$

(7)

where $u' = \mathbf{N}_u' \mathbf{u}_e / l_e$, $\theta_x' = \mathbf{N}_{\theta_x}' \mathbf{u}_e / l_e$, $I_{22} = \int_A (y^2 z^2) dA$ and $L_{p4} = \int_A (y^4 + z^4)$. The gravity element force and the fluid element force are written as

$$\mathbf{f}_g^{(e)} = \int_0^1 \mathbf{N}_u^T \rho g A l_e d\xi,$$

(8)

and

$$\mathbf{f}_f^{(e)} = \int_0^1 \left(M_f g - A_i \frac{\partial p_i}{\partial x} - \frac{1}{2} C_f \rho_f D_0 U_0^2 \right) \mathbf{N}_u^T l_e d\xi,$$

(9)

where g is the gravity acceleration, M_f is the fluid mass per unit length, A_i and A_0 are the cross-sectional areas corresponding to the inner and outer diameters of the column, D is the diameter, U_0 is the flow speed outside the column, C_f is a fluid viscous damping coefficient and p_i is the pressure inside the column (see Paidoussis et al. (2008) and Ritto et al. (2009) for details). The concentrated force at the bit is written as

$$\mathbf{f}_c = \begin{bmatrix} 0 & 0 & \cdots & -f_c & 0 \end{bmatrix}^T.$$

(10)

where f_c is the initial reaction force at the bit. The bit-rock interaction force is written as

$$\mathbf{f}_{br} = \begin{bmatrix} 0 & 0 & \cdots & f_{bit} & t_{bit} \end{bmatrix}^T.$$

(11)

in which f_{bit} is the axial force and t_{bit} is the torque about the x-axis. These two functions can be written as (Tucker and Wang, 2003)

$$f_{bit} = -\frac{\dot{u}_{bit}}{a_2 Z\left(\dot{\theta}_{bit}\right)^2} + \frac{a_3 \dot{\theta}_{bit}}{a_2 Z\left(\dot{\theta}_{bit}\right)} - \frac{a_1}{a_2},$$

$$t_{bit} = -\frac{\dot{u}_{bit} a_4 Z\left(\dot{\theta}_{bit}\right)^2}{\dot{\theta}_{bit}} - a_5 Z\left(\dot{\theta}_{bit}\right)$$

(12)

where $Z(\dot{\theta}_{bit})$ is the regularizing function such that

$$Z\left(\dot{\theta}_{bit}\right) = \frac{\dot{\theta}_{bit}}{\sqrt{\left(\dot{\theta}_{bit}\right)^2 + e^2}}.$$

(13)

In the above equation, $a_1, \ldots, a_5$ are positive constants that depend on the bit and rock characteristics as well as on the weight-on-bit (f_{bit}).

REDUCED-ORDER MODEL

To accelerate the computations of the dynamical system response, a reduced-order model is constructed. One way to construct it is to project the nonlinear dynamical equation (Eq. (1)) on a subspace $V_n \subset \mathbb{R}^m$, with $n \ll m$. In Trindade et al. (2005), the Karhunen-Loève decomposition is used to reduce a coupled axial/bending beam dynamics subjected to impacts. In the present paper, the basis used for the reduction corresponds to a basis of normal modes, which are obtained from the following generalized eigenvalue problem

$$\left(\left[K\right] + \left[K_g\left(\mathbf{u}_S\right)\right]\right)\phi = \omega^2 \left[M\right]\phi,$$

(14)

where Φ_i is the i-th normal mode and ω_i is the i-th natural frequency. Using the representation

$$\bar{\mathbf{u}}(t) = \left[\Phi\right]\mathbf{q}(t),$$

(15)

and substituting it in the equation of motion yields

$$[M][\Phi]\ddot{\mathbf{q}}(t)+[C][\Phi]\dot{\mathbf{q}}(t)+([K]+$$
$$+[K_g(\mathbf{u}_s)])[\Phi]\mathbf{q}(t)=\mathbf{g}(t)+\mathbf{f}_{br}(\ddot{\bar{\mathbf{u}}}(t)),$$

(16)

where $[\Phi]$ is a (m × n) real matrix composed by n normal modes obtained using the prestressed configuration, Eq. (14). Projecting the equation on the subspace spanned by these normal modes yields

$$[\Phi]^T[M][\Phi]\ddot{\mathbf{q}}(t)+[\Phi]^T[C][\Phi]\dot{\mathbf{q}}(t)+$$
$$+[\Phi]^T([K]+[K_g(\mathbf{u}_s)])[\Phi]\mathbf{q}(t)=$$
$$=[\Phi]^T\left(\mathbf{g}(t)+\mathbf{f}_{br}(\ddot{\bar{\mathbf{u}}}(t))\right),$$

(17)

which can be rewritten as

$$[M_r]\ddot{\mathbf{q}}(t)+[C_r]\dot{\mathbf{q}}(t)+[K_r]\mathbf{q}(t)=$$
$$=[\Phi]^T\left(\mathbf{g}(t)+\mathbf{f}_{br}(\ddot{\bar{\mathbf{u}}}(t))\right),$$

(18)

in which

$$[M_r]=[\Phi]^T[M][\Phi]$$
$$[C_r]=[\Phi]^T[C][\Phi]$$
$$[K_r]=[\Phi]^T([K]+[K_g(\mathbf{u}_s)])[\Phi]$$

(19)

are the reduced matrices. The advantage of using a reduced-order model is to end up with a diagonal reduced matrix of size 22, instead of a banded finite element matrix of size 114 (values used in the numerical simulations).

PROBABILISTIC MODEL OF THE WEIGHT-ON-HOOK

The uncertainties in the weight-on-hook are modeled using the parametric probabilistic approach; therefore, the weight-on-hook w is modeled as a random variable W. To be coherent with the physics of the problem, the probability density function of W is constructed by means of the Maximum Entropy Principle (Shannon, 1948; Jaynes, 1957a,b). This principle consists in finding the probability density function that maximizes the entropy, given some available information. To compute the distribution, an optimization problem with constraints is solved, and it is guaranteed that all available information are respected and that no unphysical conditions appear (Kapur and Kesavan, 1992).

The available information is derived from the mechanical properties of the weight-on-hook. These properties are: (1) the column must penetrate the soil, i.e., W must be lower than the weight w_2 of the column; (2) buckling must not occur, i.e., W must be greater than the buckling limit w_1; (3) the probability must go to zero when Wapproaches w_1; (4) the probability must go to zero when W approaches w_2.

Conditions (1) and (2) are expressed by setting the support of the probability density function as $]w_1,w_2[$. Conditions (3) and (4) are expressed by $E\{\ln(W-w_1)\} = \hat{c}_1$ and, $E\{\ln(w_2-W)\} = \hat{c}_2$ with $|\hat{c}_1| < +\infty$ and, $|\hat{c}_2| < +\infty$, where $E\{\ \}$ is the mathematical expectation. The reason why the logarithm, (ln), is used is because it imposes a weak decreasing of the probability density function in $w_1{}^+$ and $w_2{}^-$. To facilitate the calculus, we introduce a normalized random variable X with values in $]0,1[$, such that:

$$W = w_1\left(1 - X\right) + w_2 X$$
(20)

The expected value of W is written as

$$E\{W\} = w_1\left(1 - E\{X\}\right) + w_2 E\{X\}$$
(21)

We introduce the notation $\underline{w} = E\{W\}$ and $\underline{x} = E\{X\}$. The second moment of W may be written as

$$E\{W^2\} = E\{X^2\}\left(w_1^2 - 2w_1w_2 + w_2^2\right) +$$
$$+\underline{x}\left(-2w_1^2 + 2w_1w_2\right) + w_1^2$$
(22)

The available information is re-expressed in terms of the new random variable:

1. $X \in]0,1[$
2. $E\{\ln(X)\} = c_1$
3. $E\{\ln(1-X)\} = c_2$
(23)

with $|c_1| < +\infty$ and $|c_2| < +\infty$. The optimization problem for the maximum Entropy Principle is finally written as:

$$p_X^* = \arg\max_{p_X \in C} S\left(p_X\right)$$
(24)

where C is the space of admissible probability density functions p_X satisfying the constraints given by Eq. (23) and the entropy measure S is given by (Shannon, 1948):

$$S(p_X) = -\int_{\mathbb{R}} p_X(x) \ln(p_X(x)) \, dx$$

(25)

The probability density function, solution of the optimization problem defined by Eq. (24), is the Beta probability density function which may be written as:

$$p_X^*(x) = 1_{]0,1[}(x) \frac{\Gamma(\alpha + \beta)}{\Gamma(\alpha)\Gamma(\beta)} x^{\alpha-1} (1-x)^{\beta-1},$$

(26)

where the Gamma function $\Gamma(y) = \int_0^{+\infty} t^{y-1} \exp(-t) dt$ for y > 0. Also, α > 2 and β > 2 so that Eq. (23) holds. The random generator of independent realizations of the random variable X is already implemented in many computer codes. The mean value of X is given by

$$\underline{x} = \frac{\alpha}{\alpha + \beta},$$

(27)

and coefficient of variation is given by

$$\delta_X = \sqrt{\frac{\beta}{\alpha(\alpha + \beta + 1)}}.$$

(28)

where $\delta_X = \sigma_X / \underline{x}$, in which σ_X is the standard deviation.

The random weight-on-hook defined by Eq. (20) depends on the four parameters (w_1, w_2, α, β). For applications, w_1 and w_2 will be fixed. Parameters α and β have no physical meaning, consequently, we express them as function of the physical meaningful parameters w and δ. After some manipulations we obtain:

$$\underline{x} = \frac{\underline{w} - w_1}{w_2 - w_1},$$

(29)

$$\delta_X = $$
$$= \sqrt{\frac{\underline{w}^2(\delta^2 + 1) - (\underline{w} - w_1)^2 - 2w_1(\underline{w} - w_1) - w_1^2}{(\underline{w} - w_1)^2}}$$

(30)

and

$$\alpha = \frac{\underline{x}}{\delta^2}\left(\frac{1}{\underline{x}} - \delta_X^2 - 1\right),$$

(31)

$$\beta = \frac{\underline{x}}{\delta_X^2}\left(\frac{1}{\underline{x}} - \delta_X^2 - 1\right)\left(\frac{1}{\underline{x}} - 1\right).$$

(32)

It should be noticed that with this scheme the random weight-on-hook can be computed for any fixed values of w_1 and w_2. A specialist should be the one who gives these limits depending on the drill-string system analyzed. If the support $]w_1, w_2[$ is defined, and the mean and coefficient of variation are given, it is very simple to compute W.

STOCHASTIC DYNAMICAL SYSTEM

Using the probabilistic model of the weight-on-hook, the deterministic reduced model defined by Eq. (18) is replaced by the following stochastic equations:

$$[M_r]\ddot{\mathbf{Q}}(t) + [C_r]\dot{\mathbf{Q}}(t) + [K_r]\mathbf{Q}(t) =$$
$$= [\Phi]^T \left(\mathbf{g}(t) + \mathbf{f}_{br}\left(\dot{\mathbf{Q}}(t) + \mathbf{F}_W\right)\right),$$

(33)

where $\mathbf{Q}$ is the random response and $\mathbf{F}_W$ is a vector for which the only nonzero component is related to the axial d.o.f. of the first node $\mathbf{F}_W(1) = (W - \underline{w})$. Note that $\underline{w}$ was subtracted because the response is calculated in the prestressed configuration.

NUMERICAL RESULTS

The data used in the simulations is found in the appendix. The drill-string is discretized with 56 finite elements, and for the construction of the reduced dynamical model, 10 torsional modes, 10 axial modes and also the two rigid body modes of the structure (axial and torsional) are used, hence 22 modes in the total for the reduced model. The time integration is done using an explicit Runge-Kutta algorithm with a time step controller to keep the error within a given accuracy.

Convergence of the Stochastic Solution

Let $[U(t, s)]$ be the response of the stochastic dynamical system calculated for each realization s. The mean-square convergence analysis with respect to the number n_s of independent realizations is carried out studying the function $\mathrm{conv}(n_s)$ defined by

$$\mathrm{conv}\left(n_s\right) = \frac{1}{n_s} \sum_{j=1}^{n_s} \int_0^{t_f} \left\| \mathbf{U}\left(s_j, t\right) \right\|^2 dt.$$

(34)

Figure 2 shows that 500 simulations are sufficient to reach the mean-square convergence.

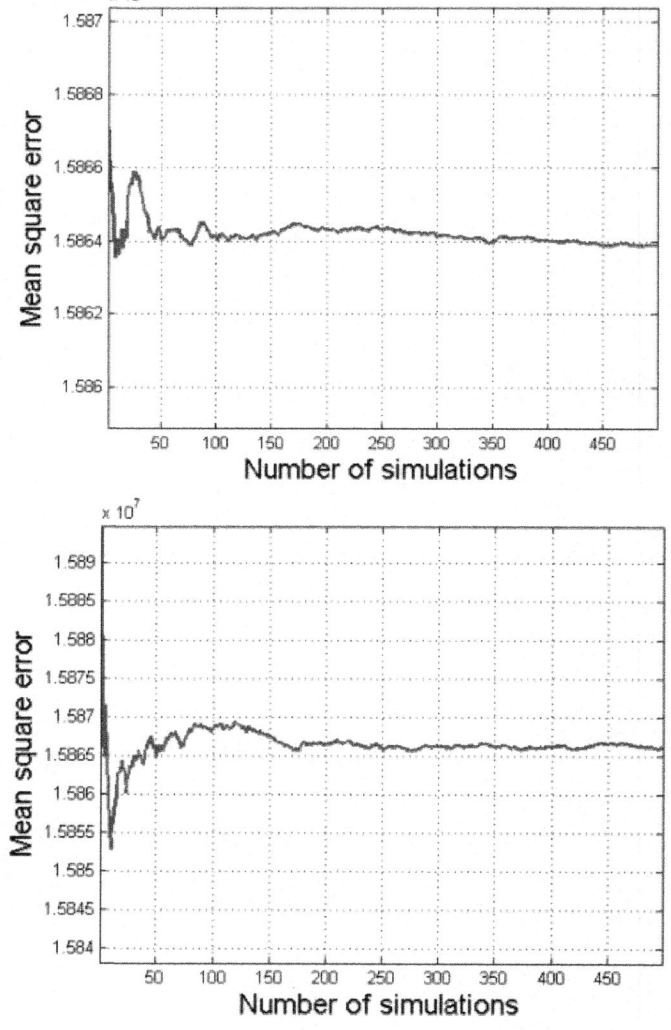

Figure 2: Mean square convergence for δ = 0.01 (top) and δ = 0.05 (bottom).

RESPONSE OF THE STOCHASTIC SYSTEM

The stochastic system's response is analyzed in this Section. Note that the dispersion of the response is all due to the random weight-on-hook, which is modeled as shown in Section four (Probabilistic model of the weight-on-hook). We may identify experimentally the parameters of the probabilistic model of the random weight-on-hook (mean and coefficient of variation) directly by measuring the weight-on-hook, or indirectly by measuring the dynamical response of the bit, for instance. In both cases, the identification procedure can be done using, for example, the Maximum Likelihood method (Aldrich, 1997; Spall, 2005). As we know that there are other sources of uncertainties for the problem, the experimentally identification procedure should be performed considering all the modeled random variables.

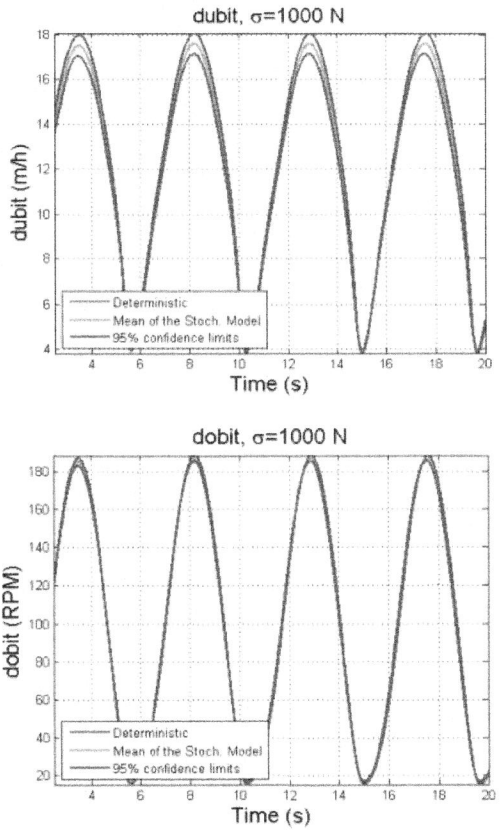

Figure 3: 95% envelope for σ = 1000 N. Rate-of-penetration, ROP (top); and rotational speed of the bit (bottom).

Figure 3 shows the 95% envelope (that is to say the confidence region constructed with a probability level of 0.95) for the rate-of-penetration and the rotational speed of the bit for a standard deviation σ = 1000 N, which means δ = $\sigma/\underline{w}$ is approximately 1×10^{-3}. The envelopes (the upper and lower envelopes of the confidence region) are calculated using the method of quantiles, Serfling (1980).

We are plotting two important variables: the rate-of-penetration (ROP) and the rotational speed at the bit (ω_{bit}). So, we analyze the influence of the random weight-on-hook in the system response. It can be seen that, for σ = 1000 N, the random response presents tie confidence intervals. Figure 4 shows the stochastic response of the torque and force on the bit.

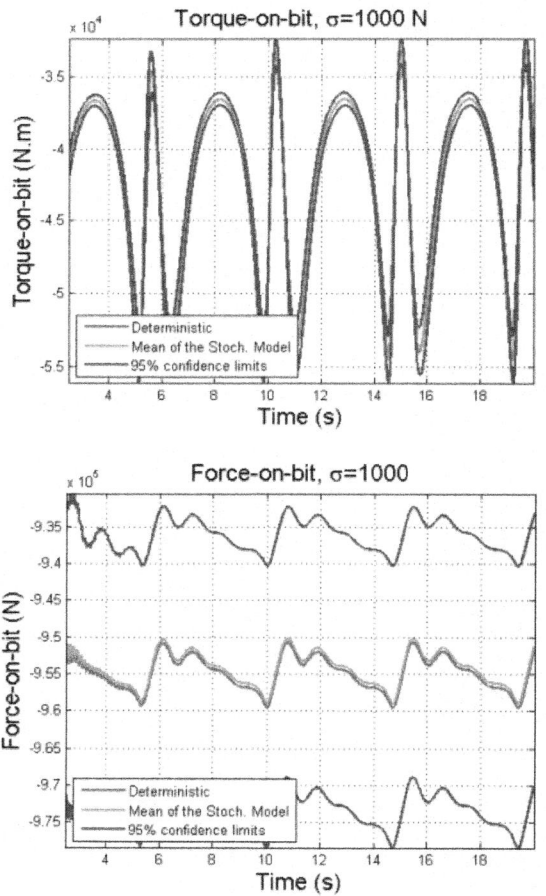

Figure 4: 95% envelope for σ = 1000 N. Torque-on-bit (top); and force-onbit (bottom).

It is noted that for $\sigma = 1000$ N, the response changes just a little, therefore, σ will be increased in the next analysis. In our analysis we cannot increase σ too much, because the model used for the bit-rock interaction assumes a weight-on-bit $f_{bit} \sim -100$ kN. Hence, the standard deviation σ of the W is increased in a way that the f_{bit} has a maximum variation around 5%, that is to say that σ_{max} = 5000 N and, therefore, $\delta_{max} \sim 0.05$ (0.5% variation), which is a constraint to our analysis. But, as it will be seen, a small variation on W may cause a big variation in the system response. Figures 5 and 6 show the system response for $\sigma = 3000$ N ($\delta \sim 0.003$).

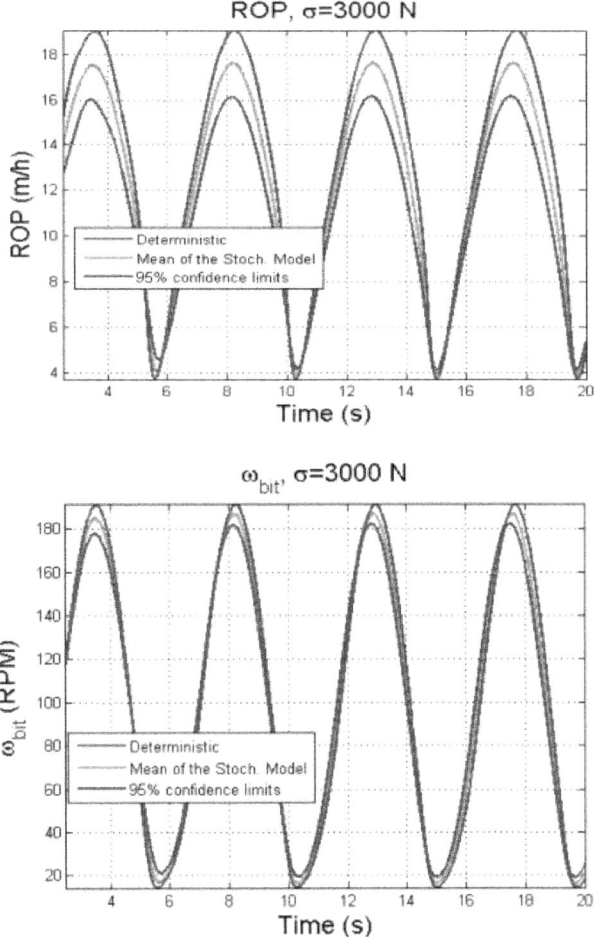

Figure 5: 95% envelope for $\sigma = 3000$ N. Rate-of-penetration, ROP (top); rotational speed of the bit (bottom).

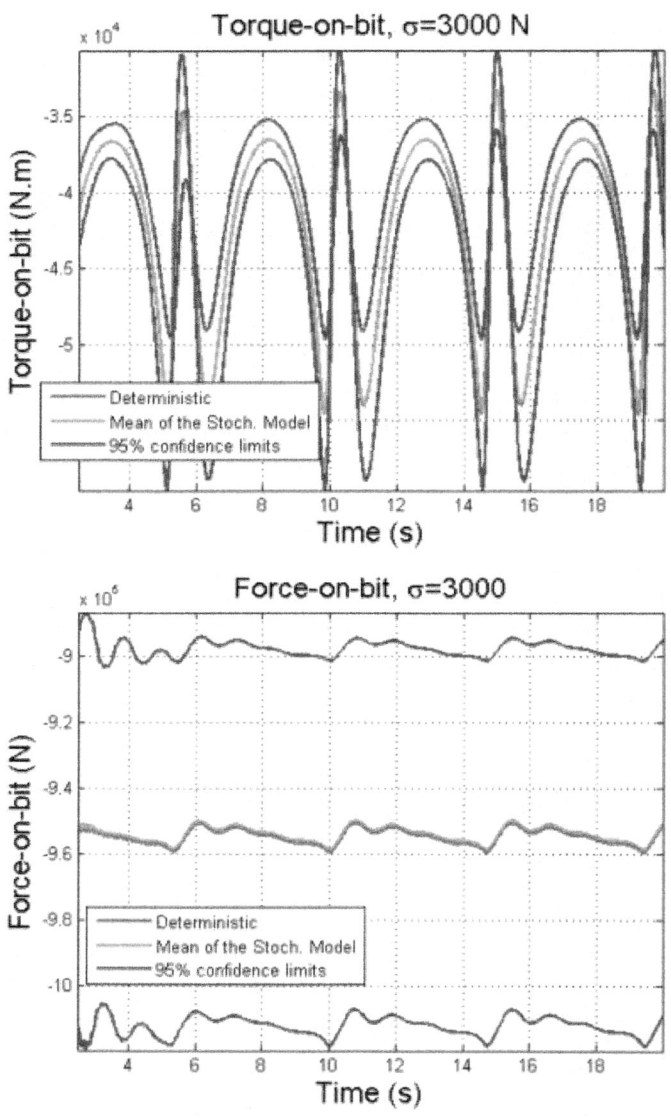

Figure 6: 95% envelope for σ = 3000 N. Torque-on-bit (top); and force-onbit (bottom).

We want to see how uncertainties in the weight-on-hook affect the performance of the system; hence, Fig. 7 shows the evolution of the dispersion of the response for four dynamic responses: ROP, rotational speed of the bit, torque-on-bit, and force-on-bit. The dispersion of the response is calculated taking the square root of the variance divided by the value of the mean response for each time instant (it is the instant coefficient of variation). It can be noticed that the mean coefficient of variation of the force-on-bit (FOB) (see Fig. 7(d)) is about 3%, which is much more than 0.3%, which is the coefficient of variation of the random weight-on-hook W. This might be explained due to the fact that the absolute value of W is much higher than the absolute value of FOB; therefore, a small percentage variation of W has a great effect in the FOB, in terms of percent. What is more critical is the fact that the dispersion is even higher for the ROP and the rotational speed of the bit (Figs. 7(a) and (b)).

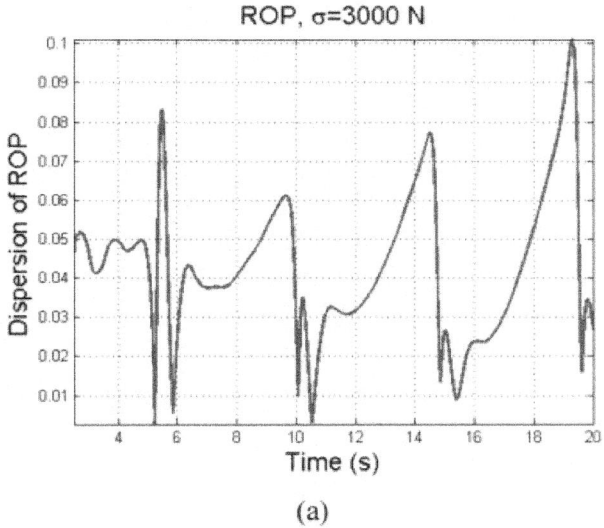

(a)

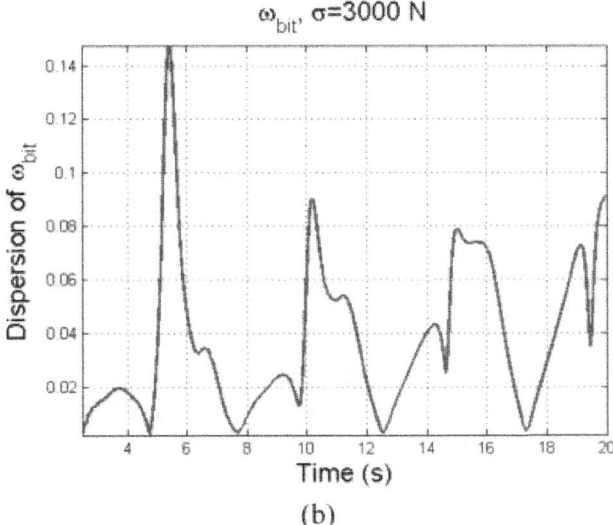

(b)

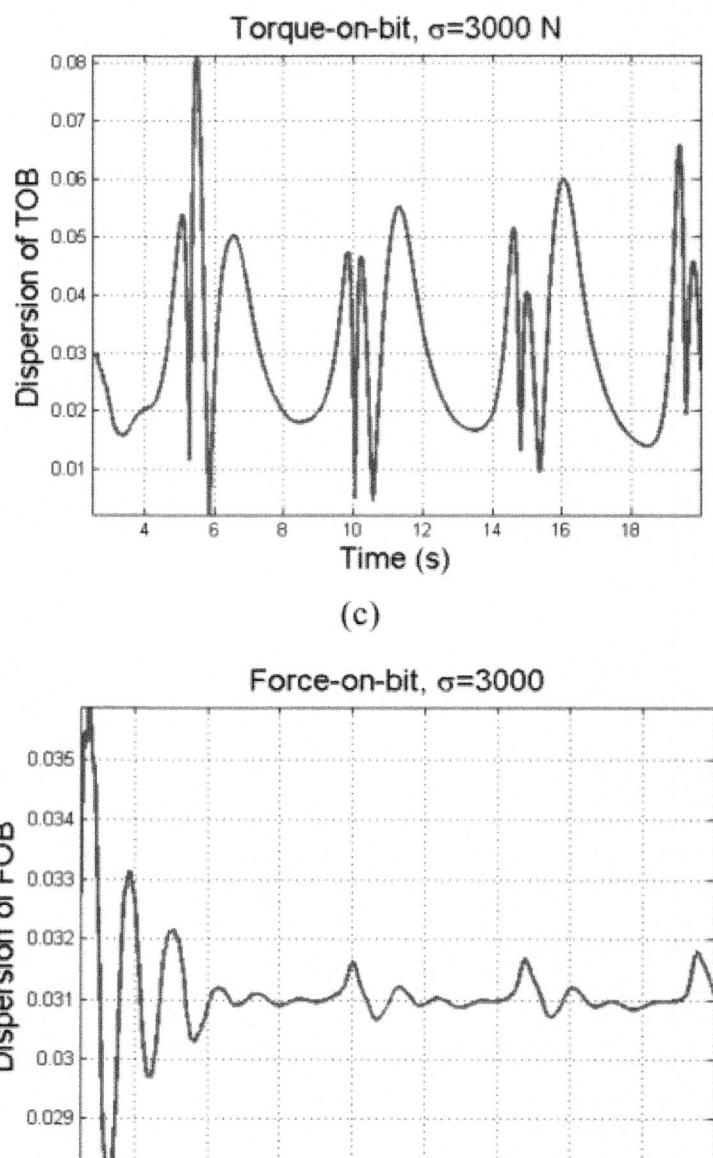

Figure 7: Dispersion of the response for σ = 3000 N. (a) Rate-ofpenetration, ROP; (b) rotational speed of the bit; (c) torque-on-bit; and (d) force-on-bit.

Figure 8 shows the system response for σ = 5000 N(δ ~ 0.005).

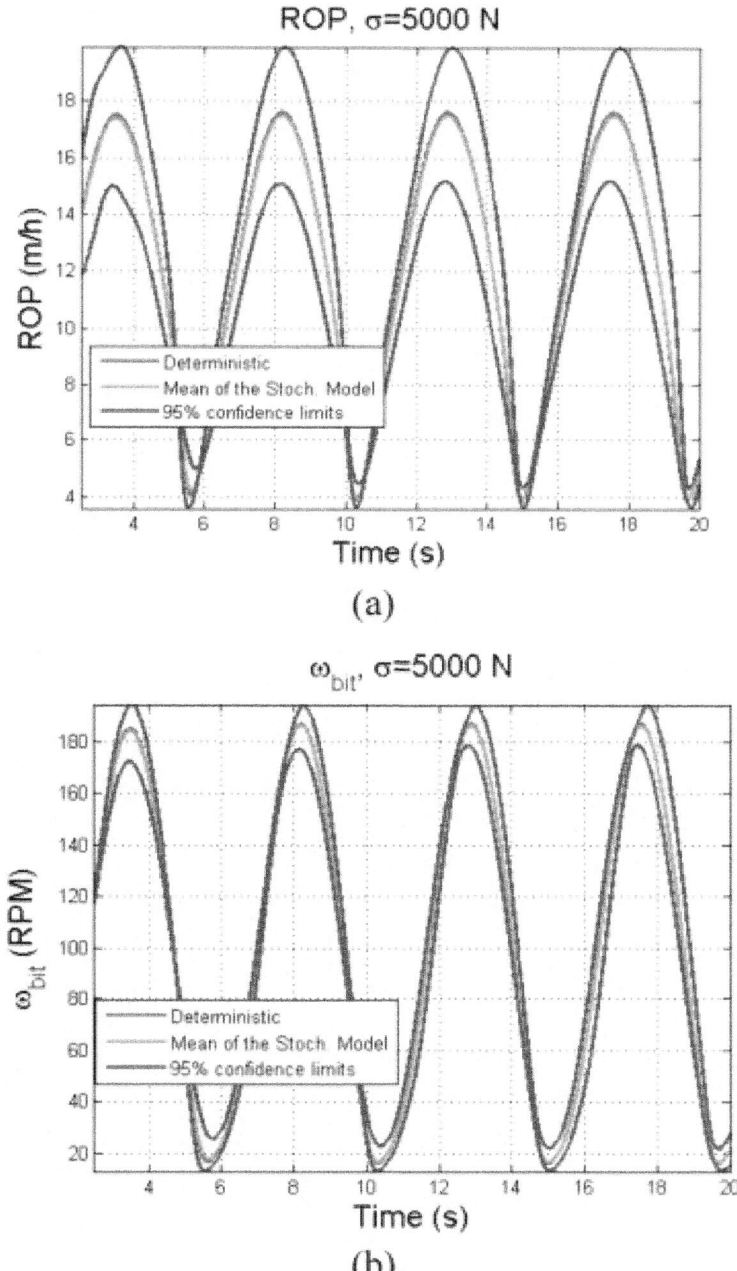

(a)

(b)

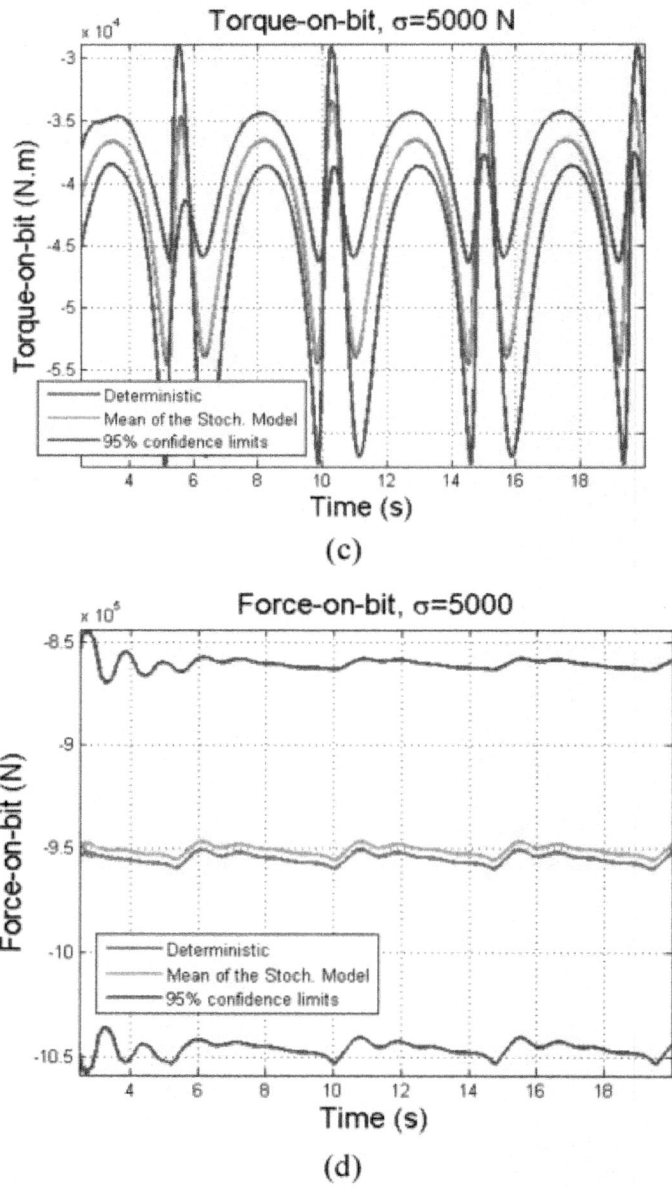

Figure 8: 95% envelope for σ = 5000 N. (a) Rate-of-penetration, ROP; (b) rotational speed of the bit; (c) torque-on-bit; and (d) force-on-bit.

As expected, as δ increases the envelope of the response gets wider. Figure 9 shows the dispersion of the response for σ = 5000 N of the W.

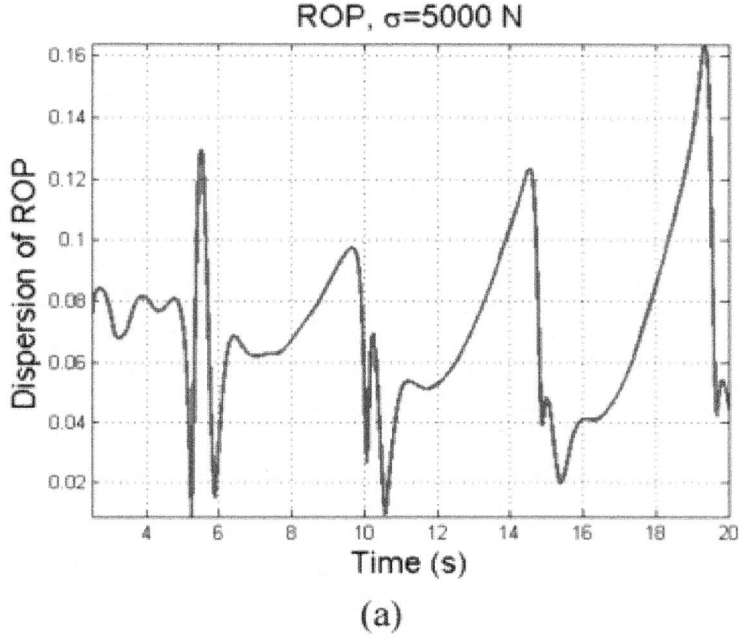

(a)

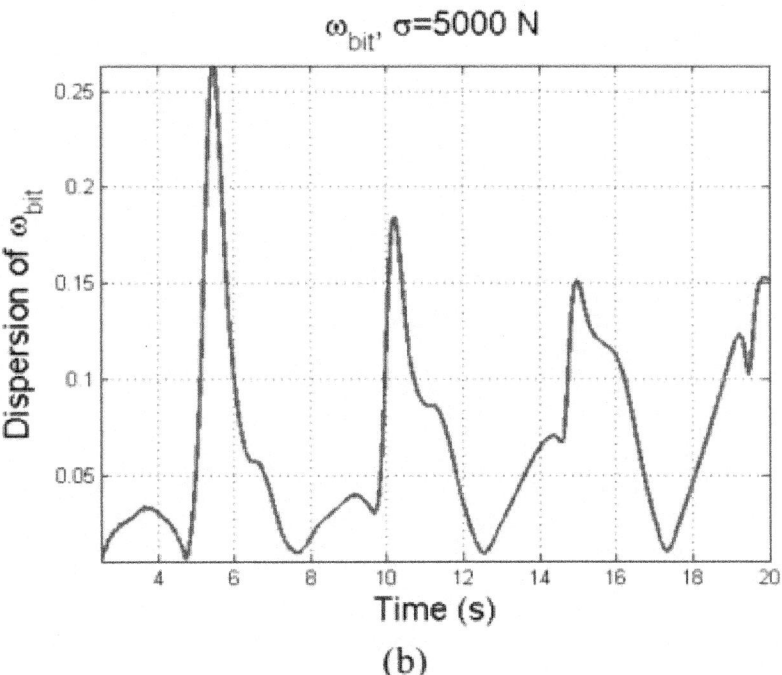

(b)

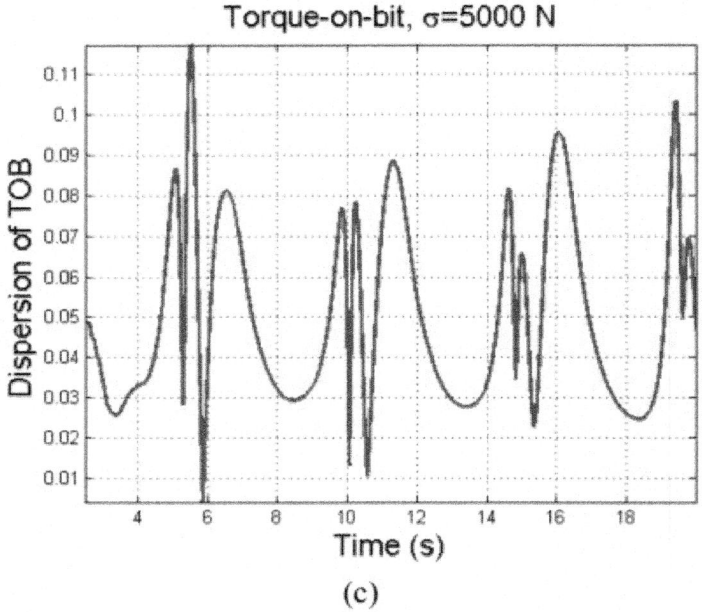

(c)

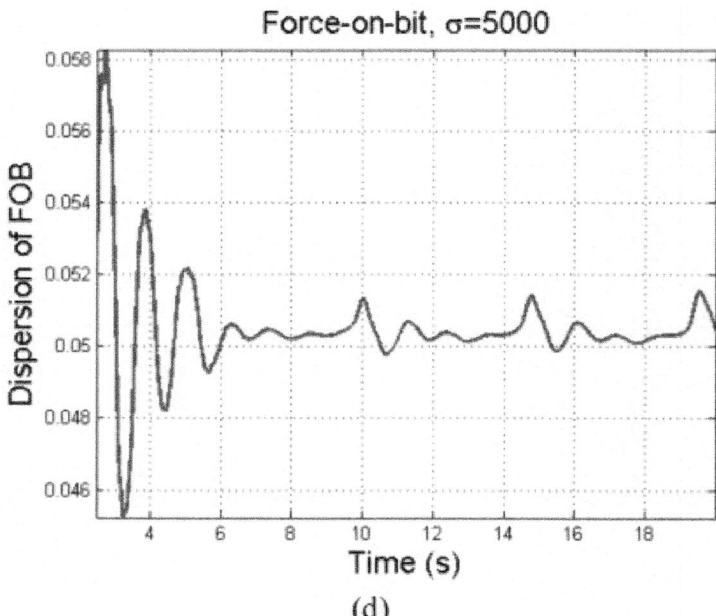

(d)

Figure 9: Dispersion of the response for σ = 5000 N. (a) Rate-ofpenetration, ROP; (b) rotational speed of the bit; (c) torque-on-bit; and (d) force-on-bit.

It is noted that, even for a small variation of W (~ 0.5%), there is a big dispersion in the response. See, for instance, the rate-of-penetration: the mean coefficient of variation is 4.3%, which is more than eight times greater than the coefficient of variation of W. It gets worse if we take the maximum coefficient of variation, which is 16%. It means that if the W has a coefficient of variation of half percent, the variation in the ROP may achieve sixteen percent and the coefficient of variation of the rotational speed of the bit may achieve twenty six percent!

CONCLUDING REMARKS

A stochastic model of the drill-string dynamics has been analyzed. The weight-on-hook has been modeled as a random variable with probability density functions constructed using the Maximum Entropy Principle. It has been shown that the system response is sensitive to dispersion on the weight-on-hook. There are many sources of uncertainties in this problem; hence, more stochastic analysis should be done to identify the uncertainties that most affect the performance of the system.

ACKNOWLEDGEMENTS

The authors acknowledge the financial support of the Brazilian agencies CNPQ, CAPES, and FAPERJ.

APPENDIX A - DATA USED IN THE SIMULATIONS

$\underline{w}$ = 1.06×10^6 N (mean weight-on-hook),

w_1 = 0.49×10^6 N (weight-on-hook related to the buckling limit),

w_2 = 1.16×10^6 N (weight of the column),

L_{dp} = 1400 m (length of the drill pipe),

L_{dc} = 200 m (length of the drill collar),

D_{odp} = 0.127 m (outside diameter of the drill pipe),

D_{odc} = 0.2286 m (outside diameter of the drill collar),

D_{idp} = 0.095 m (inside diameter of the drill pipe),

D_{idc} = 0.0762 m (inside diameter of the drill collar),

E = 210 GPa (elasticity modulus of the drill string material),

ρ = 7850 kg/m^3 (density of the drill string material),

Ω_x = 100 RPM (constant speed at the top),

ρ_f = 1200 kg/m^3 (density of the fluid),

C_f = 1.25×10-2 (fluid viscous damping coefficient),

f_c = 100 kN (initial reaction force at the bit),

g = 9.81 m/s^2 (gravity acceleration),

a_1 = 3.429×10^{-3} m/s (constant of the bit-rock interaction model),

a_2 = $5.672 \times 10_{-8}$ m/(N.s) (constant of the bit-rock interaction model),

a_3 = 1.374×10^{-4} m/rd (constant of the bit-rock interaction model),

a_4 = 9.537×10^{-6} N.rd (constant of the bit-rock interaction model),

a_5 = $1.475e \times 10^3$ N.m (constant of the bit-rock interaction model),

e = 2 rd/s (regularization parameter).

The damping matrix is constructed using the relation $[C] = a[M] + b([K] + [K_g(\mathbf{u}_S)])$ with $a = 0.01$ and $b = 0.0003$.

REFERENCES

1. Aldrich, J., 1997, "R.A. Fisher and the making of maximum likelihood 1912-1922", Statistical Science, Vol. 12, No. 3, pp. 162-176.

2. Jaynes, E., 1957a, "Information theory and statistical mechanics", The Physical Review, Vol. 106, No. 4, pp. 1620-630.

3. Jaynes, E., 1957b, "Information theory and statistical mechanics II", The Physical Review, Vol. 108, pp. 171-190.

4. Kapur, J.N., Kesavan, H.K., 1992, "Entropy Optimization Principles with Applications", Academic Press, Inc., USA.

5. Khulief, Y.A., AL-Naser, H., 2005, "Finite element dynamic analysis of drillstrings", Finite Elements in Analysis and Design, Vol. 41, pp. 1270-1288.

6. Kotsonis, S.J., Spanos, P.D., 1997, "Chaotic and random whirling motion of drillstrings", Journal of Energy Resources Technology, Vol. 119, No. 4, pp. 217-222.

7. Paidoussis, M.P., Luu, T.P., Prabhakar, S., 2008, "Dynamics of a long tubular cantilever conveying fluid downwards, which then flows upwards around the cantilever as a confined annular flow", Journal of Fluids and Structures, Vol. 24, pp. 111-128.

8. Piovan, M.T., Sampaio, R., 2009, "Modelos continuos de sondas de perforación para la industria petrolera: Análisis de enfoques y su discretización", Revista Internacional de Métodos Numéricos para Cálculo y Diseño en Ingeniería, Vol. 35, No. 3, pp. 259-277.

9. Ritto, T.G., Sampaio, R., Cataldo, E., 2008, "Timoshenko beam with uncertainty on the boundary conditions",Journal of the Brazilian Society of Mechanical Sciences and Engineering, Vol. 30, No. 4, pp. 295-303.

10. Ritto, T.G., Soize, C., Sampaio, R., 2009, "Nonlinear dynamics of a drillstring with uncertain model of the bit-rock interaction", International Journal of Non-Linear Mechanics, Vol. 44, No. 8, pp. 865-876.

11. Ritto, T.G., Soize, C., Sampaio, R., 2010, "Robust optimization of the rate of penetration of a drill-string using a stochastic nonlinear dynamical model", Computational Mechanics, article in press, DOI: 10.1007/s00466-009-0462-8.

12. Rubinstein, R.Y., 2007, "Simulation and the Monte Carlo Method", 2nd Edition, Series in Probability and Statistics, John Wiley and Sons, New Jersey, USA.

13. Sampaio, R., Piovan, M.T., Lozano, G.V., 2007, "Coupled axial/torsional vibrations of drilling-strings by mean of nonlinear model", Mechanics Research Comunications, Vol. 34, No. 5-6, pp. 497-502.

14. Sampaio, R., Ritto, T.G., 2008, "Short course on dynamics of flexible structures - deterministic and stochastic analysis", Seminar on Uncertainty Quantification and Stochastic Modeling, PUC-Rio, September.

15. Serfling, R.J., 1980, "Approximation Theorems of Mathematical Statistics", John Wiley and Sons, USA.

16. Shannon, C.E., 1948, "A mathematical theory of communication", Bell System Tech. J., Vol. 27, pp. 379-423 and 623-659.

17. Soize, C., 2000, "A nonparametric model of random uncertainties for reduced matrix models in structural dynamics", Probabilistic Engineering Mechanics, Vol. 15, pp. 277-294.

18. Spall, J.C., 2005, "Introduction to Stochastic Search and Optimization", John Wiley and Sons, Hoboken, NJ, USA.

19. Spanos, P., Politis, N., Esteva, M., Payne, M., 2009, "Drillstring vibrations", Advanced Drilling and Well Technology (Society of Petroleum Engineers), pp. 117-156.

20. Spanos, P.D., Payne, M., Secora, C., 1997, "Bottom-hole assembly modeling and dynamic response determination", Journal of Energy Resources Technology, Vol. 119, No. 3, pp. 153-158.

21. Trindade, M.A., Wolter, C., Sampaio, R., 2005, "Karhunen-Loève decomposition of coupled axial/bending of beams subjected to impacts", Journal of Sound and Vibration, Vol. 279, pp. 1015-1036.

22. Tucker, R.W., Wang, C., 1999, "An integrated model for drill-string dynamics", Journal of Sound and Vibration, Vol. 224, No. 1, pp. 123-165.

23. Tucker, R.W., Wang, C., 2003, "Torsional vibration control and Cosserat dynamics of a drill-rig assembly", Meccanica, Vol. 38, No. 1, pp. 143-159.

Chapter 11

DYNAMICS AND CONTROL OF HIGH-RISE BUILDINGS UNDER MULTIDIRECTIONAL WIND LOADS

Aly Mousaad Aly,[1] Alberto Zasso,[2] and Ferruccio Resta[2]

[1]Department of Mechanical Engineering, Faculty of Engineering, Alexandria University, Alexandria 21544, Egypt

[2]Department of Mechanical Engineering, Politecnico di Milano, Via G. La Masa 1, 20156 Milano, Italy

ABSTRACT

This paper presents a procedure for the response prediction and reduction in high-rise buildings under multidirectional wind loads. The procedure is applied to a very slender tall building that is instructive. The structure is exposed to both cross-wind and along-wind loads obtained from pressure measurements on a rigid model (scaled 1 : 100) that was tested in a wind tunnel with two different configurations of the surroundings. In the theoretical formulation, dynamic equations of the structure are introduced by finite element and 3D lumped mass modeling. The lateral responses of the building in the two directions are controlled at the same time using tuned mass dampers (TMDs) and active tuned mass dampers (ATMDs) commanded by LQR and fuzzy logic controllers, while the effects of the uncontrolled torsional response of the structure are simultaneously considered. Besides their simplicity, fuzzy logic controllers showed similar trend as LQR controllers under multidirectional wind loads. Nevertheless, the procedure presented in this study can help decision makers, involved in the design process, to choose among innovative solutions like structural control, different damping techniques, modifying geometry, or even changing materials.

INTRODUCTION

Civil engineering structures are an integral part of our modern society. Traditionally, these structures are designed to resist static loads. However, they may be subjected to dynamic loads like earthquakes, winds, waves, and traffic. Such loads can cause severe and/or sustained vibratory motion, which can be detrimental to the structure and human occupants. Because of this, safer and more efficient designs are sought out to balance safety issues with the reality of limited resources. Wind-induced vibrations in buildings are of increasing importance, as the use of high-strength, lightweight materials, longer floor spans, and more flexible framing systems results in structures that are more prone to vibrations. In tall buildings, wind-induced vibrations may cause annoyance to the occupants (especially in the upper floors), impaired function of instruments, or structural damage.

Traditionally, wind-induced response of tall buildings in the along-wind direction are evaluated using some codes and formulas [1–4]. However, these standards provide little guidance for the critical cross-wind and torsional responses. This is due to the fact that the cross-wind and torsional responses result mainly from the aerodynamic pressure fluctuations in the separated shear layers and the wake flow fields, which made it difficult to have an acceptable direct analytical relation to the oncoming velocity fluctuations [5, 6]. In addition, these methods may have some limitations, especially for accounting of surrounding tall buildings. Moreover, responses are restricted to some few modes, and the process of evaluating such response depends on much assumption. On the other hand, wind tunnel pressure measurements and finite element modeling (FEM) of the structures are an effective alternative for determining these responses. Wind tunnel tests have been industry wide accepted reliable tools for estimating wind loads on tall buildings. For tall buildings, there are two types of testing: (1) high-frequency base balance (HFBB) and (2) high-frequency pressure integration (HFPI). Inherent in the HFFB approach is the fact that only the global wind loads at the base of the test model are known. The test results from the HFBB measurements can be analyzed using frequency-domain or time-domain techniques to get the building responses. The frequency domain approach has been dominant over time domain approach for its lesser requirement of computational power though it involves more approximations compared to the time-domain approach. Nevertheless, with the current technology where computational power is no longer a problem, the time-domain method is becoming a popular analysis technique. The time-domain method allows determination of wind responses directly from the equation of motion using the measured time history, thereby avoiding all the simplifying assumptions used in the frequency domain technique [7]. However, even if the more accurate time-domain approach is used for the analysis of the response, the three-dimensional (3D) mode shapes found in complex tall buildings complicate the use of the HFBB test results for predicting the response [8, 9]. In general, mode shape correction factors for the HFBB technique are necessary for the assessment of wind-induced responses of a tall building. This is to account for the significant uncertainties in the prediction of generalized forces due to the nonideal mode shapes as well as presumed wind loading distributions [10, 11]. HFPI with the time-domain approach can be more accurate provided that enough coverage of pressure taps on the model's outer surface is performed.

HFPI technique is based on simultaneous pressure measurements at several locations on a building's outer surface. Pressure data can be used for the design of the claddings as well as the estimation of the overall design loads. The HFPI technique cancels out any inertial effects that may be included in the overall loads measured by the base balance if the HFBB technique was used. Time histories of wind forces at several levels of tall building models can be obtained in the wind tunnel with a multichannel pressure scanning system. This enables the building responses to be computed directly in the time domain for buildings with simple or complex mode shapes. Finite element models (FEMs) can be used for describing the dynamic behavior of the structures. HFPI with FEM have the advantages of considering complex shapes of structures with nonuniform mass distribution and can easily account for any required number of mode shapes to be considered in the response analysis.

Preliminary analysis of tall buildings in their preliminary design stages help the designer to make decision by modifying the design or adding passive, active, or semiactive control techniques. Structural control has recently been the subject of much discussion among structural designers. Structural control can potentially provide safer and more efficient structures. The concept of employing structural control to minimize structural vibrations was proposed in the 1970s [12]. The purpose of structural control is to absorb and to reflect the energy introduced by dynamic loads. The reduction of structural vibrations occurs by adding a mechanical system that is installed in a structure. The control of structural vibrations can also be done by various means, such as modifying rigidities, masses, damping, or shape, and by providing passive or active counter forces [13,14]. Passive, active and semiactive [15, 16] types of control strategies have been proposed.

McNamara [17] studied the tuned mass damper (TMD) as an energy-absorbing system to reduce wind-induced structural response of buildings in the elastic range of behavior. Active control techniques are studied intensively for the control of the response of tall buildings under wind loads [18–22]. The most commonly used active control device for tall buildings is the active tuned mass damper (ATMD). TMDs and ATMDs are shown to be effectiveness in the response reduction of tall buildings under wind loads [23–28].

The aim of this study is to present practical procedure for the response prediction and reduction in a very slender high-rise building under multidirectional wind loads. The procedure is schematically presented in Figure 1. Wind loads were obtained from an HFPI experiment conducted in a wind tunnel. The tower responses in the two lateral directions combined with the torsional responses (effect of higher modes on the responses is studied) are evaluated. Two important voids associated with procedures to aid in the design are considered: the first is on the distributions of the wind loads; the second is on the effects of the higher modes. Consideration of these two problems needs wind tunnel pressure measurements on the surface of the building and FEM. The building is modeled using the finite element techniques and a 3D lumped mass model. The uncontrolled responses obtained using the two techniques of modeling are compared. Active control of the structure using LQR and fuzzy logic controllers under wind that is attacking from different directions is proposed. In this study, the lateral responses of the building in the two directions are controlled at the same time, while the effects of the uncontrolled torsional responses of the structure are simultaneously considered.

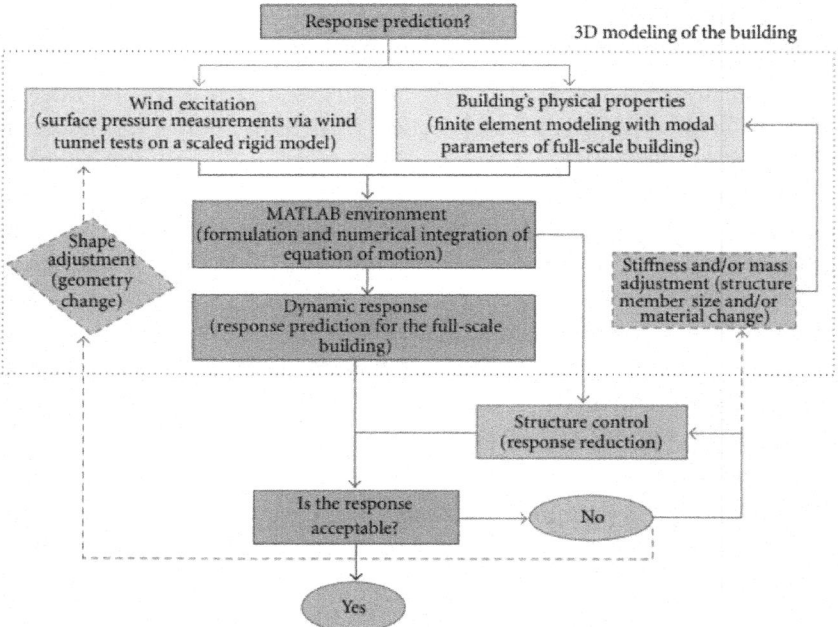

Figure 1: Schematic representation of the proposed procedure for response prediction and reduction in tall buildings under wind loads.

3D MODELING OF THE BUILDING

A 48-strory steel tower proposed in Aly et al. [23] is used in this research. The FEM of the tower, along with the coordinate system, is shown in Figure 2. The full-scale building has a height of about 209 m and a rectangular cross section of B/D ≈ 3 (B: chord length and D: thickness). The aspect ratio in the y-direction is about 11, which makes it very sensitive to strong winds. Modal parameters of the FEM for the first six modes are given in Table 1.

Table 1: Modal parameters of the FEM model

Mode number*	Generalized mass × 10^7 (kg·m²)	Generalized stiffness × 10^9 (N·m)	Frequency (Hz)	Modal damping
1	1.2953	0.0147	0.1694	0.0102
2	0.9937	0.0178	0.2132	0.0112
3	0.4945	0.0222	0.3370	0.0150
4	0.8724	0.1115	0.5689	0.0234
5	0.8273	0.2153	0.8120	0.0326
6	0.3544	0.1600	1.0695	0.0426

*Modes 1 and 4 are lateral displacements in x-direction; modes 2 and 5 are lateral disp. in y-direction while modes 3 and 6 are torsion.

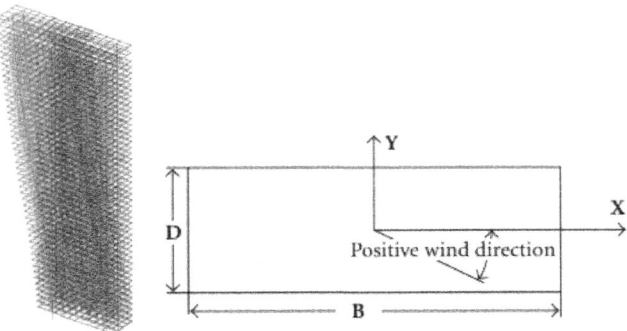

Figure 2: Finite element model with the coordinate system and wind direction (m and m).

Equations of Motion in Modal Form

Equations of motion governing the behavior of the structure under wind loads are

$$M\ddot{X} + C\dot{X} + KX = F(t), \tag{1}$$

where X = [x y] ^{T} is a 2n × 1 vector and n is the number of nodes, while x and y are vectors of nodal displacements in x and y directions, respectively. F(t) = [F_x(t) Fy(t)]^T, in which F_x(t) and Fy (t) are n × 1 vectors of external forces acting in x and y directions, respectively. Using the first N modes obtained by FEM with the next transformation

$$\mathbf{X} = \mathbf{\Phi Q},$$
(2)

where $\mathbf{\Phi}$ is $2n \times N$ matrix of eigenvectors and Q is $N \times 1$ vector of generalized displacements; that is,

$$\mathbf{\Phi} = \begin{bmatrix} \phi_1(x_1) & \phi_2(x_1) & \cdots & \phi_N(x_1) \\ \phi_1(x_2) & \phi_2(x_2) & \cdots & \phi_N(x_2) \\ \vdots & \vdots & & \vdots \\ \phi_1(x_n) & \phi_2(x_n) & \cdots & \phi_N(x_n) \\ \phi_1(y_1) & \phi_2(y_1) & \cdots & \phi_N(y_1) \\ \phi_1(y_2) & \phi_2(y_2) & \cdots & \phi_N(y_2) \\ \vdots & \vdots & & \vdots \\ \phi_1(y_n) & \phi_2(y_n) & \cdots & \phi_N(y_n) \end{bmatrix},$$

$$\mathbf{Q} = \begin{Bmatrix} q_1 \\ q_2 \\ \vdots \\ \vdots \\ q_N \end{Bmatrix}.$$
(3)

Substituting by (2) into (1) and premultiplying by $\mathbf{\Phi}^T$, one obtains

$$\mathbf{\Phi}^T \mathbf{M} \mathbf{\Phi} \ddot{\mathbf{Q}} + \mathbf{\Phi}^T \mathbf{C} \mathbf{\Phi} \dot{\mathbf{Q}} + \mathbf{\Phi}^T \mathbf{K} \mathbf{\Phi} \mathbf{Q} = \mathbf{\Phi}^T \mathbf{F}(t).$$
(4)

By assuming the damping matrix, C, to be proportional damping, (4) results into six uncoupled equations

$$m_{11}\ddot{q}_1 + c_{11}\dot{q}_1 + k_{11}q_1$$

$$= \sum_{i=1}^{n} \Phi_1(x_i)F_{x,i}(t) + \sum_{i=1}^{n} \Phi_1(y_i)F_{y,i}(t) = GF_1,$$

$$m_{22}\ddot{q}_2 + c_{22}\dot{q}_2 + k_{22}q_2$$

$$= \sum_{i=1}^{n} \Phi_2(x_i)F_{x,i}(t) + \sum_{i=1}^{n} \Phi_2(y_i)F_{y,i}(t) = GF_2,$$

$$\vdots$$

$$m_{NN}\ddot{q}_N + c_{NN}\dot{q}_N + k_{NN}q_N$$

$$= \sum_{i=1}^{n} \Phi_N(x_i)F_{x,i}(t) + \sum_{i=1}^{n} \Phi_N(y_i)F_{y,i}(t) = GF_N,$$
(5)

where m_{ii}, c_{ii}, k_{ii}, and G_{Fi} are generalized mass, generalized damping, generalized stiffness, and generalized force of the ith mode, respectively. Using the measurements obtained by the pressure transducers, pressure coefficients (matrix C_p) are evaluated at each tap location as a function of time. These values are used with the full-scale model to give the pressure distribution on the surface. The pressure values on the surface of the prototype can be calculated as

$$\mathbf{P}(\text{space, time}) = \frac{1}{2}\rho U^2 \mathbf{C}_P(\text{space, time}), \tag{6}$$

where P(space,time) is a matrix containing the pressure values on the surface of the full scale model as a function of space (x, y, and z) and time; ρ is the air density which is assumed to be 1.25 kg/m3 (according to [2]), and U is the prototype mean wind speed. The wind load at any node of the outer surface is the integration of the pressure over the surface area in the vicinity of the node as

$$\mathbf{F}(\text{nodes, time}) = \int \mathbf{P}(\text{space, time})dA. \tag{7}$$

This means that once the time history of the pressures on the outer surfaces is calculated, the external forces acting on the nodes of the surface can be computed. The excitation forces acting on the internal nodes are of course equal to zero. The $q_{i(t)}$ are then solved from each of (5). SIMULINK is used for the numerical solution of these equations [29].

3D Lumped Mass Model

For control purposes, a 3D lumped mass model is derived from the original FEM. In this model, the total mass of the building was assumed to be lumped at the positions of the floors, and it was assumed for the floors to perform a general 3D movement (each floor has two translations in the x and y directions in addition to the torsional rotation). The building alone (without the control devices) is modeled dynamically using a total of 144 degree-of-freedom. In general, the equations of motion for an -story building moving in both the two transverse directions and in torsion are written as

$$\mathbf{M}_s\ddot{\mathbf{x}} + \mathbf{C}_s\dot{\mathbf{x}} + \mathbf{K}_s\mathbf{x} = -\mathbf{F} + \Lambda\mathbf{f}, \tag{8}$$

where x = [X Y Θ] $_T$. The terms X = [x_1 x_2 $\cdots$ x_n] and Y = [y_1 y_2 $\cdots$ y_n] are row vectors of the displacements of the centre of mass of each floor in the x and y directions, respectively, and Θ = [θ_1 θ_2 $\cdots$ θ_n] is the vector of the rotations of each floor about the vertical axis (z-axis) whilen is the number of stories. The mass matrix, M_s, and the stiffness matrix, K_s, have the following form:

$$
M_s = \begin{bmatrix} M & 0 & 0 \\ 0 & M & 0 \\ 0 & 0 & I \end{bmatrix}, \qquad K_s = \begin{bmatrix} K_x & 0 & 0 \\ 0 & K_y & 0 \\ 0 & 0 & K_\theta \end{bmatrix},
$$

$$(9)$$

here $M = \mathrm{diag}([m_1 \ m_2 \ \cdots m_n])$ is a diagonal matrix of lumped masses, $I = \mathrm{diag}([I_1 \ I_2 \ \cdots \ I_n])$ in which Ii is the moment of inertia of the ith floor, K_x, K_y, and K_θ are the stiffness matrices in the transverse directions (x and y) and the torsional direction, respectively. The stiffness matrix of the spatial model (3D lumped mass model) is obtained by assuming the stiffness between floors as a combination of cantilever and shear rigidities. MATLAB codes (MATLAB 2008; [29]) were written and used to derive the best stiffness matrix that gives the closest mode shapes to those of the FEM and the same first six natural frequencies. In (8), the disturbance $F = [F_x \ F_y \ T]^T$ is a vector of excitation in which F_x and F_y are two vectors of the horizontal loads acting in the x and y directions, respectively, and T is a vector of the external torsional wind loads. Also, f is the vector of control forces, where its coefficient matrix Λ is the matrix determined by the location of control devices.

Wind loading vectors (F_x, F_y, and T) lumped at the position of floors are obtained from wind tunnel pressure tests conducted at the wind tunnel of Politecnico di Milano [30] on a scaled 1 : 100 rigid model of the tower. Such large-scale allows for the advantage of testing the model at high Reynolds number with minimum blockage due to the huge dimensions of the test section. Pressure taps were distributed on the outer surface of the test model. To allow for sufficient pressure measurements (see [7]), 400 taps were mapped on the outer surface of the model. Pressure taps were distributed to cover the entire outer surface with more intense at the upper part of the test model (see Figure 5(a)). Pressure data were acquired at a frequency of 62.5 Hz using 448 pressure taps. Figure 6 shows a typical spectrum of measured pressure data at model scale. Pressure data were integrated on the outer surface of the building (see Figure 5(b)) to obtain the corresponding time history of the two directional wind loads at each floor in addition to torsion. For the estimation of the wind loads at each floor, the tributary area for each floor was gridded into smaller areas and the time history of the wind loads at each area was found by using the Cp records of the closest pressure tap (see Figure 7). Codes were written in MATLAB to estimate the time histories of the wind forces acting at the center of each smaller area. After that, the floor forces in the two directions were obtained from the summation of the forces in each lateral direction. The torsion at each floor was the resultant of the summation of the force moments about the floor center. The surrounding buildings within a radius of 500 m from the centre of the tower were also scaled 1 : 100 to be presented on the turning test table according to the type of the test configuration used. The wind profile represents a typical urban terrain as shown in Figure 3 [31]. The reference mean wind speed (U_{ref} = 15 m/s) was measured at a height of 1 m. Prototype reference mean wind speed is assumed to be 30 m/s. The target for the wind profiles is the Eurocode 1 [2]. The turbulence intensities in the longitudinal, lateral, and vertical directions are referred to by I_u, I_v, and I_w, respectively. Two different test configurations with the same wind profile are considered in this paper (see Figure 4). In the first configuration, the building is subjected to the wind load without the existence of other tall buildings, and is referred to as Config. no. 1. In the second configuration, the rigid model of the building was tested with the existence of all the surrounding buildings. This configuration is referred to as Config. no. 2. Further details about the wind tunnel experiment are given in Aly [24].

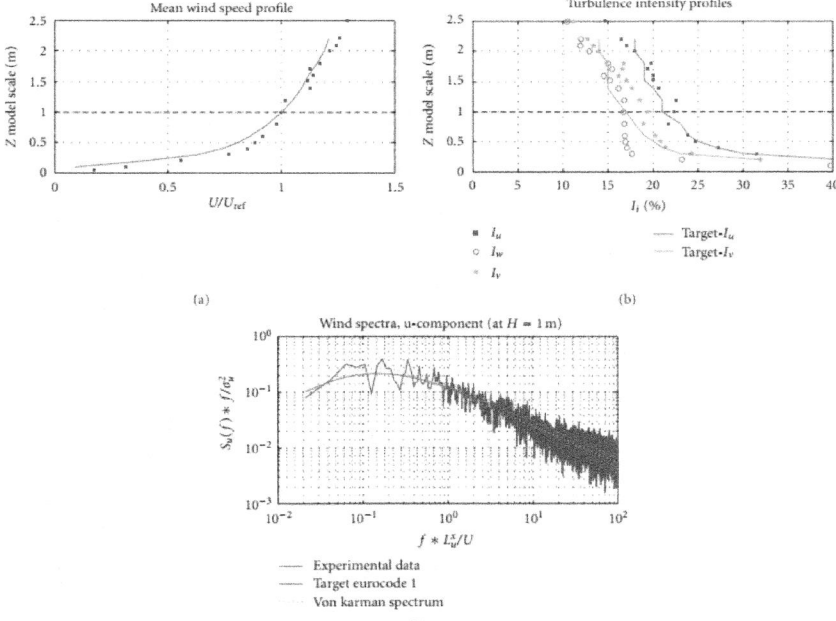

Figure 3: Mean wind speed profile, turbulence intensity profiles, and wind spectra (is the integral scale).

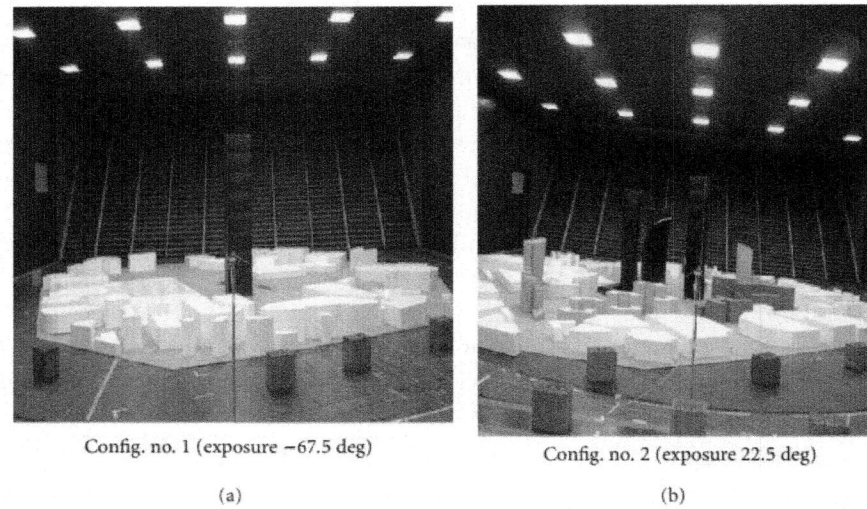

Config. no. 1 (exposure −67.5 deg)

(a)

Config. no. 2 (exposure 22.5 deg)

(b)

Figure 4: Two different configurations are used.

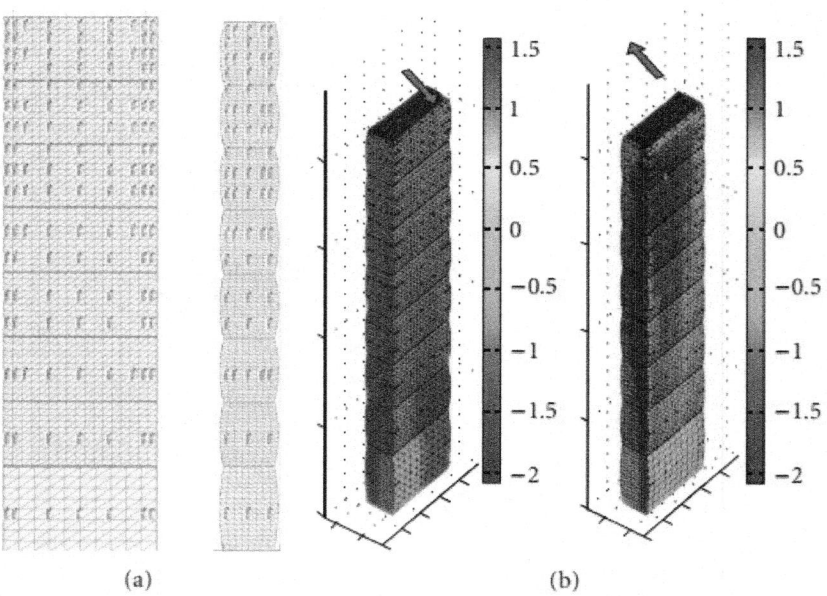

(a)

(b)

Figure 5: Pressures on the outer surface of a scaled 1 : 100 model were obtained from a wind tunnel test: (a) pressure tap distribution (elevation and side view), (b) mean surface pressure coefficient distribution (for 292.5 deg).

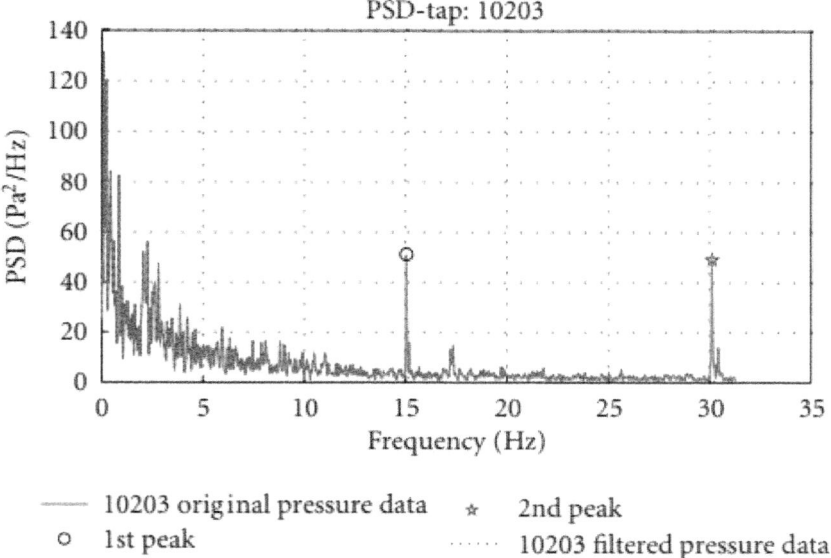

Figure 6: Typical spectrum of measured pressure data (at model scale).

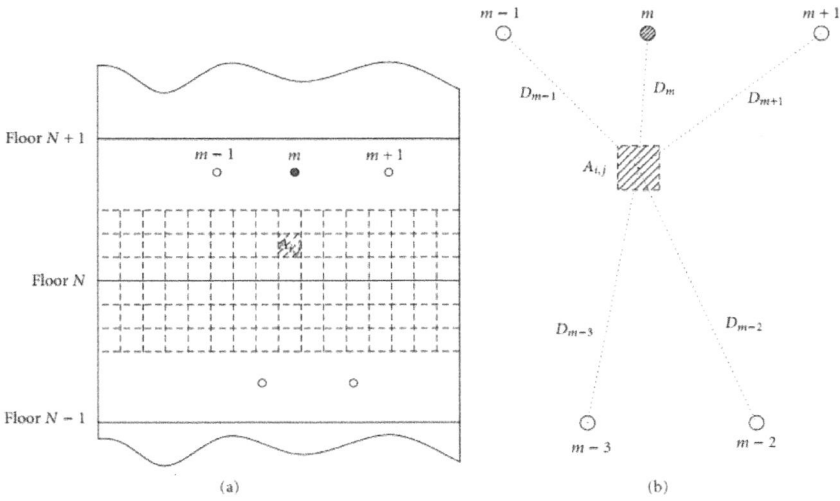

Figure 7: Wind load estimation from pressure data: the tributary area of floor was divided into smaller areas; pressure forces acting on each smaller area, , were calculated based on pressure data at the nearest pressure tap, .

Due to the fact that the building's mass is symmetrical, and the study is based on the assumption that the structure is responding in the linear region, the lateral and the torsional behavior of the building may be studied alone, then the response time histories may be combined simultaneously. In this study, the plane motion of the structure in the xdirection is controlled using both the TMDs and ATMDs. However, due to the fact that the control of the response in the x-direction will not affect the response of the building in the y-direction, another TMD and ATMD are designed to control the lateral in-plane response in the y-direction. Following that, the uncontrolled torsional response is added simultaneously to the two lateral responses to give the overall response in the two lateral directions.

The state reduction approach derived by Davison [32] and summarized later in Wu et al. [33] is used in this study (also see Lu et al. [20]). In this approach, the 48 degree-of-freedom (DOF) in-plane system is reduced to 15 DOF, where the first 30 modes are retained [23]. Note that the condition for this approach was that the response in terms of displacements and accelerations of the 15 DOF and 48 DOF are very much the same (see Section 4). This model is referred to as the reduced order system (ROS). The addition of the TMD increases the DOF to 16. The system with the TMD is referred to as ROS-TMD. In a similar task, ROS-ATMD refers to the ROS utilizing ATMD.

The state equation of the ROS that corresponds to the full order system (FOS) in (8) can be expressed as

$$\dot{z} = Az + Bf + Ew, \tag{10}$$

in which $z = [\bar{X}; \dot{\bar{X}}]$ is the 32-dimensional state vector, $\overline{X}$ is a vector of the in-plane displacements of floors 3, 6, 9, 12, 16, 19, 22, 26, 29, 32, 35, 38, 41, 44, and 48 in addition to the displacement of the inertial mass of the damper. A is a (32 × 32) system matrix, B is a 32 location vector, and E is a 32 excitation vector. In the system reduction above, the wind loads acting on each of the 15 floors above are computed from the wind loads F acting on each of the 48 floors by lumping wind forces on adjacent floors at the locations that correspond to the 15 DOF model. The controlled output vector, yc, and the measured output, ym, of the ROS described by (10) can be expressed as

$$y_c = C_c z + D_c f + F_c w_x,$$

$$y_m = C_m z + D_m f + F_m w_x + \nu, \tag{11}$$

where C_c, D_c, F_c, C_m, D_m, and F_m are matrices with appropriate dimensions and v is the measurement noise vector. The model used for controller design was further reduced as follows:

$$\dot{z}_r = A_r z_r + B_r f + E_r w_x,$$

$$y_{cr} = C_{cr} z_r + D_{cr} f + F_{cr} w_x,$$

$$y_{mr} = C_{mr} z_r + D_{mr} f + F_{mr} w_x + v_r, \tag{12}$$

where z_r is a 6-dimensional state vector of the reduced order system, y_{cr} is a controlled output vector identical of y_c defined by (11), ymr is the measured output vector; v_r is the measurement noise, and C_{cr}, D_{cr}, F_{cr}, C_{mr}, D_{mr}, and F_{mr} are appropriate matrices.

CONTROLLERS AND LIMITATIONS

In this study, both TMDs and ATMDs are used for the reduction of the lateral responses of the building. However, in order to make the design of such control systems more realistic and applicable, the following restrictions and assumption are applied.(i) The mass of the TMD in the x-direction is 100 ton, while the mass of the TMD in the y-direction is 150 ton. Such restrictions are applied to avoid excessive weight on the roof (the overall mass on the roof is about 0.625% of the overall building's mass)..(iii)The maximum stroke of the actuators is restricted to 1.5 m.(iv)The maximum control force of the actuator in the -direction is restricted to 100 kN, and that in the -direction is restricted to 25 kN.(v)The computational delay and the sampling rate of the digital controller are 0.001 s.(vi)Three acceleration measurements are available for each lateral direction (at floor 30, roof, and mass of the TMD).

Note that the tower required a TMD with heavier mass and ATMD with higher control force in one lateral direction than the other, which was basically attributed to geometry. A linear-quadratic regulator (LQR) design with output weighting is selected to give the desired control force using the MATLAB function (lqry.m). The state-feedback law $f = -G_{zr}$ minimizes the cost function

$$J(f) = \int_0^\infty (y'_{cr} Q y_{cr} + f' R f) dt, \tag{13}$$

where G is the feedback gain matrix, zr is a 6-dimensional state vector of the reduced order system, y_{cr} is the measured output vector, the symbol (') denotes transpose, and Q and R are weighting matrices. Parametric studies were performed with various weighting matrices Q, corresponding to various regulated output vectors. The results of these parametric studies indicated that an effective controller could be designed by selecting a vector of regulated responses to include the velocities of each floor.

For comparison reasons, fuzzy logic controllers are used in this study to command the actuators of the ATMDs (see Nguyen et al. [34]). From a design point of view, fuzzy logic controllers do not require the complexity of a traditional control system. The measured accelerations can be used directly as input to the fuzzy controller. The main advantages of using a fuzzy control algorithm are summarized in Battaini et al. [35] and Samali et al. [36]. According to Samali et al. [36], uncertainties of input data are treated in a much easier way by fuzzy control theory than by classical control theory. Since fuzzy controllers are based on linguistic synthesis, they possess inherent robustness. Fuzzy controllers can be easily implemented in a fuzzy chip with immediate reaction time and autonomous power supply. Furthermore, the design of fuzzy controller does not require state reduction or concerning about observers. Only two acceleration measurements were used (floor 30 and roof).

The input variables to the fuzzy controller were selected as accelerations of floors 30 and 48 and the output as the control force. The membership functions for the inputs were defined and selected as seven triangles with overlaps as shown in Figure 8. For the output, they were defined and selected as nine triangles with overlaps as shown in Figure 9. The fuzzy variables used to define the fuzzy space are ZR (zero), PVS (positive very small), PS (positive small), PM (positive medium), PL (positive large), PVL (positive very large), NVS (negative very small), NS (negative small), NM (negative medium), NL (negative large), and NVL (negative very large). The rule-base for computing the desired current is presented in Table 2 [36].

Table 2: Control rule base [36]

Acceleration of 48th floor	Acceleration of 30th floor						
	NL	NM	NS	ZR	PS	PM	PL
NL	PVL	PVL	PL	PVS	ZR	ZR	ZR
NM	PL	PL	PM	PVS	ZR	ZR	ZR
NS	ZR	NVS	PM	PS	PVS	ZR	ZR
ZR	ZR	ZR	NVS	ZR	PVS	ZR	ZR
PS	ZR	ZR	NVS	NS	NM	PVS	ZR
PM	ZR	ZR	ZR	NVS	NM	NL	NL
PL	ZR	ZR	ZR	NVS	NL	NVL	NVL

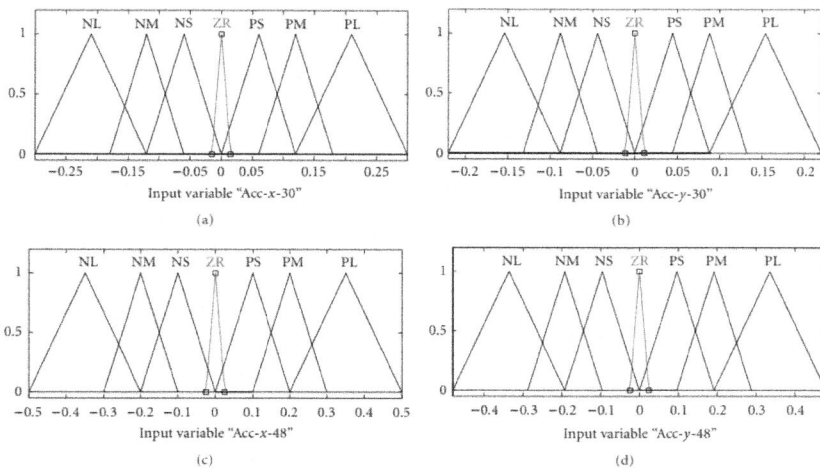

Figure 8: Membership functions for the input measured accelerations in the -direction (Acc--30, Acc--48) and the -direction (Acc--30, Acc--48).

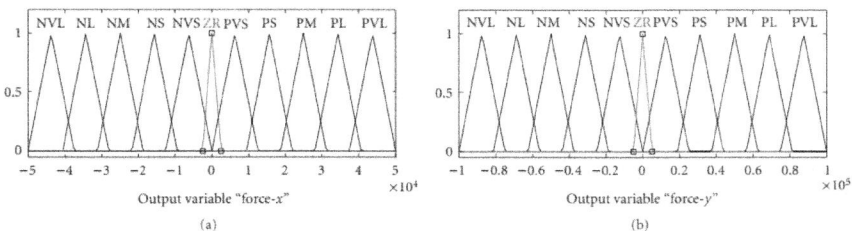

Figure 9: Membership functions for the output control force in the -direction (Force-) and the -direction (Force-).

RESULTS AND DISCUSSION

Table 3 gives the response of the top corner of the building in the -direction for an incident angle of 0° under different consideration of mode shapes. It is shown that the displacement response of this building is dominated by the first lateral mode in the -direction (modes 1 : 2 in the table). Nevertheless, this underestimates the displacement response by 3% to 4.4% and the acceleration response by about 12% to 17%. Note that the aspect ratio of this building in the -direction is about 11. This means that for very slender buildings, the solo consideration of the first lateral mode may lead to significant error in the estimation of the response, especially for the acceleration response. Table 4 lists the response of the top corner of the tower in the -direction for an incident angle of 90° under different consideration of mode shapes. It is shown that the displacement and acceleration response are dominated by the first lateral mode in the -direction (modes 1 in the table). Note that the aspect ratio of this building in the -direction is about 3.6. This means that for buildings with low aspect ratio, the solo consideration of the first lateral mode may be sufficient for the estimation of the response. Figure 10 shows the power spectra of the acceleration response of the top corner of the building in the two lateral directions. The figure shows that the third mode (torsion) contributes significantly to the acceleration in the -direction. In general, results given by the figure, Tables 3 and 4 show that the responses of tall buildings under winds are dominated by the first few modes (for this specific building, the first two lateral modes and the first torsional mode can be sufficient).

Table 3: Response of the top corner of the tower in the -direction for an incident angle of 0°

Mode	RMS-disp. (m)	Max-disp. (m)	RMS-accel. (m/s²)	Max-accel. (m/s²)
1	0.000 (−100%)	0.001 (−99.8%)	0.000 (100%)	0.001 (−99.9%)
1 : 2	0.129 (−4.4%)	0.587 (−2.8%)	0.199 (−17.1%)	0.855 (−11.8%)
1 : 3	0.136 (0.7%)	0.613 (1.5%)	0.238 (−0.8%)	0.980 (1.1%)
1 : 4	0.136 (0.7%)	0.613 (1.5%)	0.238 (−0.8%)	0.980 (1.1%)
1 : 5	0.135 (0%)	0.606 (0.3%)	0.239 (−0.4%)	0.966 (−0.3%)
1 : 6	0.135 (0%)	0.604 (0%)	0.240 (0%)	0.969 (0%)

Table 4: Response of the top corner of the tower in the -direction for an incident angle of 90°

Mode	RMS-disp. (m)	Max-disp. (m)	RMS-accel. (m/s²)	Max-accel. (m/s²)
1	0.188 (1.1%)	0.646 (−0.5%)	0.203 (−0.5%)	0.654 (−3.5%)
1 : 2	0.188 (1.1%)	0.646 (−0.5%)	0.203 (−0.5%)	0.654 (−3.5%)
1 : 3	0.187 (0.5%)	0.648 (−0.2%)	0.204 (0%)	0.653 (−3.7%)
1 : 4	0.186 (0%)	0.649 (0%)	0.204 (0%)	0.676 (−0.3%)
1 : 5	0.186 (0%)	0.649 (0%)	0.204 (0%)	0.676 (−0.3%)
1 : 6	0.186 (0%)	0.649 (0%)	0.204 (0%)	0.678 (0%)

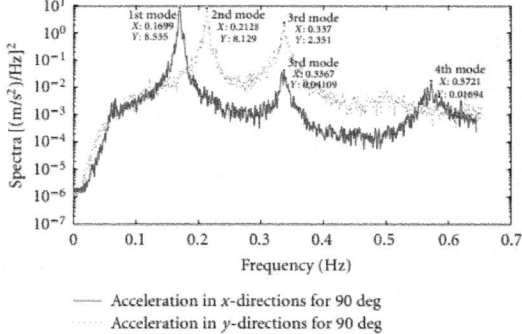

—— Acceleration in x-directions for 90 deg
········ Acceleration in y-directions for 90 deg

Figure 10: Power spectra of the acceleration response of the top corner of the building in the two lateral directions.

Figure 11 gives displacement and acceleration responses of a point at the top corner of the building for the FEM, the 3D full order system (3D-FOS), and the 3D reduced order system (3D-ROS). The figure shows that the response in terms of displacements and accelerations of the three types of modeling are very much the same (see Section 4). This means that FE modeling, 3D lumped mass modeling, and 3D reduced order modeling of tall buildings under wind loads can give an accurate assessment of the response provided that the first dominant modes are retained. The figure shows also that the cross-wind response is higher than the along-wind response. This reveals the importance of the procedure proposed in this study as many design codes and formula may provide accurate estimate of the along-wind response but less guidance for the estimation of the critical cross-wind and torsional response. The results show that the building is very much vulnerable to wind loads. This may be due to its low weight and its low dominant frequencies.

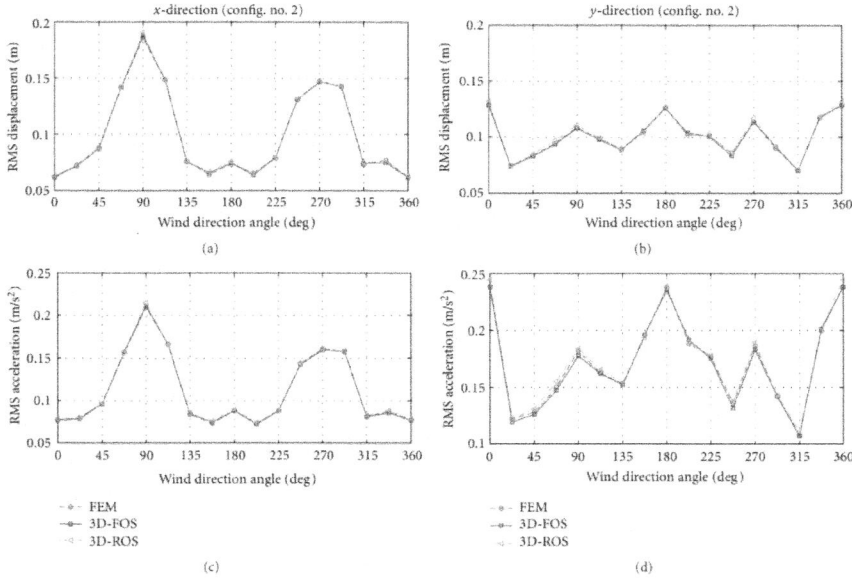

Figure 11: displacement and acceleration responses of a point at the top corner for FEM, 3D full order system (3D-FOS), and 3D reduced order system (3D-ROS).

The building required a TMD with heavier mass and ATMD with higher control force in one lateral direction than the other. This may be attributed to geometry. Figures 11–14 show the controlled and uncontrolled responses of the tower under wind loads for two test configurations at different incident angles. Two examples of control are considered, TMDs and ATMDs with LQR and fuzzy logic controllers. For each example, the controlled responses in the and directions are plotted with the uncontrolled responses. The controlled and uncontrolled responses of the tower are evaluated by simulations [29]. Four evaluation criteria are used to examine the performance of the proposed controllers. Evaluation criteria include: rms displacements, maximum displacements, rms accelerations, and maximum accelerations of the top corner of the tower in the two lateral directions. The figures are superimposed by ellipses indicating the position of the most unfavourable responses (uncontrolled, with TMDs, with ATMDs (LQR), and with ATMDs (fuzzy)) over the two configurations in both and directions. The amount of reduction in the highest response achieved by TMDs and ATMDs over the worst uncontrolled response is indicated in the figures.

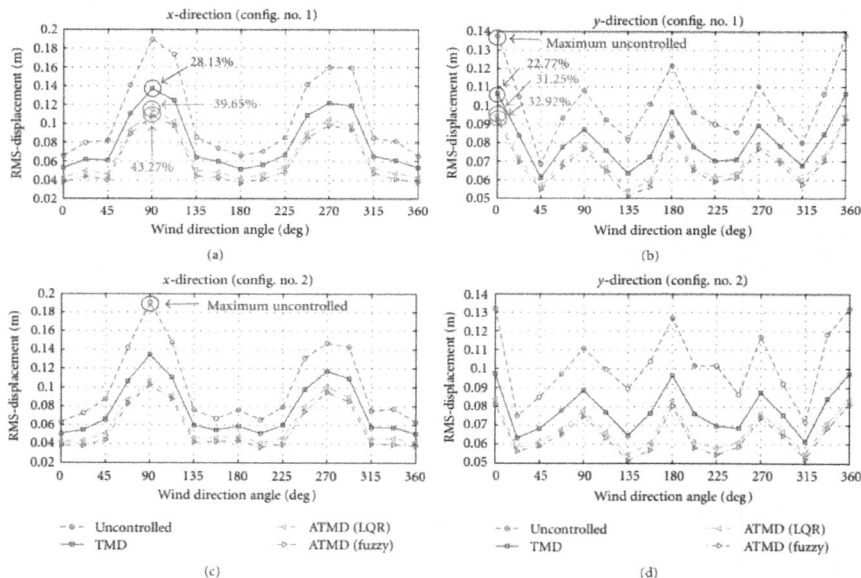

Figure 12: RMS-displacements of the top corner of the tower.

Figures 11 and 12 give controlled and uncontrolled rms displacements and max displacements of the top corner of the tower in both the and directions. It is shown that TMDs have a great effect on the reduction of the displacement response of the building. Reductions achieved by TMDs in the displacements responses range from 22–30% over the worst uncontrolled response. Generally, TMDs are able to give good reduction in the rms displacements in both the and directions for all wind incident angles. Reductions achieved by ATMDs in the displacement responses range from 29%–43% over the worst uncontrolled response. ATMDs with fuzzy logic controllers are able to enhance the reduction in the displacement responses over LQR (perhaps Q and R could be more optimal) most of the time (by about 1% to 5%). They also have a general similar trend over all of the wind attack angles.

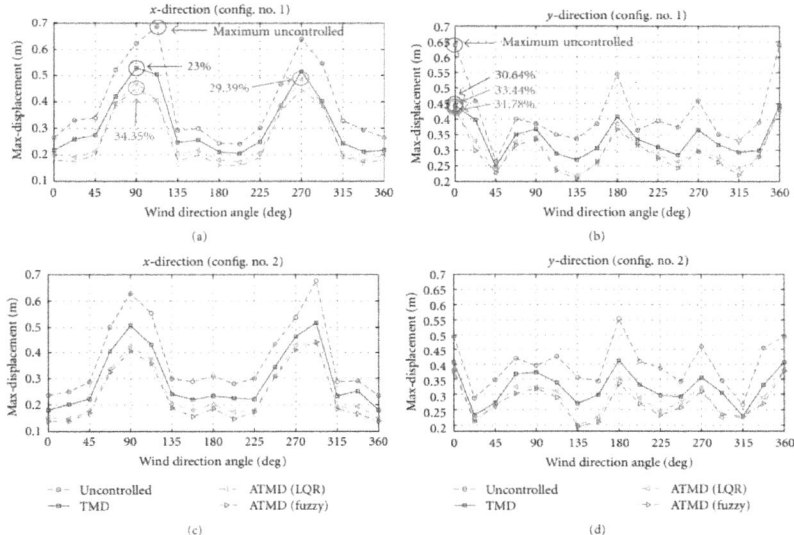

Figure 13: Maximum displacements of the top corner of the tower.

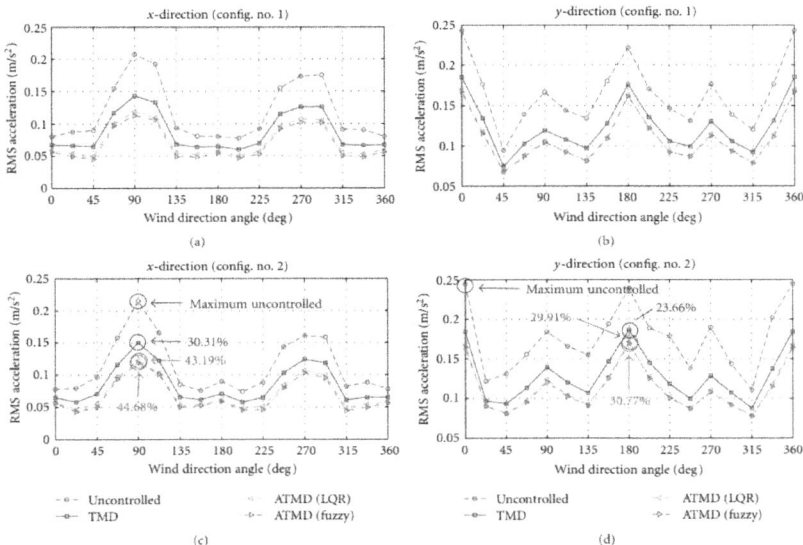

Figure 14: RMS accelerations of the top corner of the tower.

Figures 13 and 14 give controlled and uncontrolled rms accelerations and maximum accelerations (Figure 15) of the top corner of the tower in both the and directions. It is shown that the TMDs have a significant effect on the reduction of the acceleration response of the building. Reductions achieved by TMDs in the acceleration responses range from 16%–30% over the worst uncontrolled response.

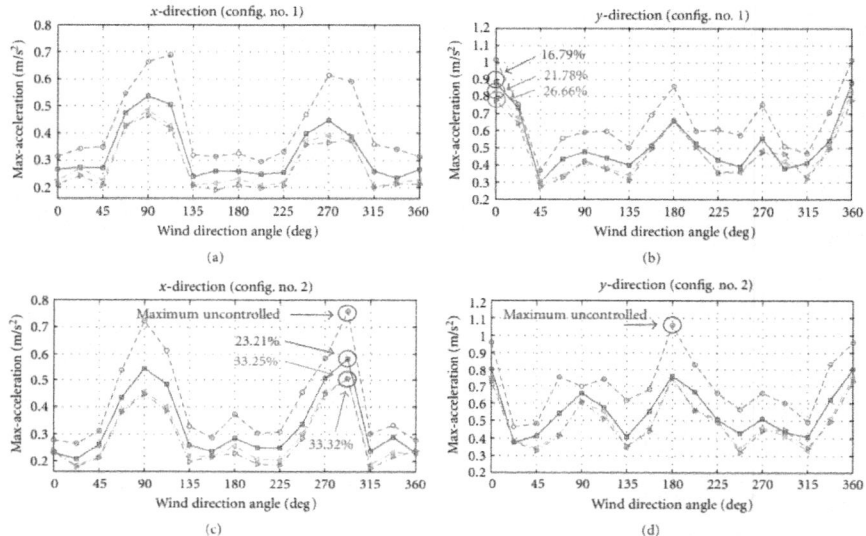

Figure 15: Maximum accelerations of the top corner of the tower.

Generally, TMDs are able to give good reduction in the rms displacements in both the and directions for all the wind incident angles. However, the performance is limited in reducing the along-wind maximum acceleration of the tower in the -direction under Config. no. 2, when the wind direction angle is 90°. This may be due to the interference effects of two high-rise buildings in the oncoming wind (see Figure 4). Results also show that ATMDs are able to enhance the reduction in the responses. Reductions achieved by ATMDs in the acceleration responses range from 21%–43% over the worst uncontrolled response. ATMDs with fuzzy logic controllers are able to enhance the reduction in the acceleration responses over LQR, and in general, they have a similar trend over all of the wind incident angles.

As a general comment on Figures 11–14, one can see that the performance of the controllers is much better in the -direction. In addition, the capability of the controllers to reduce the responses (especially maximum accelerations at angles 0° and 180°) in the -direction is limited. This may be due to the effect of vortex shedding on the across-wind responses. Moreover, the structure is slender in -direction (see Figure 2). The structure is also stiffer in the -direction (see Table 1 for natural frequencies). However, the procedure presented in this study permits the response of tall buildings to be assessed and controlled in the preliminary design stages which can help decision makers, involved in the design process, to choose among innovative design solutions like structural control, considering several damping techniques, modifying geometry, or even changing materials (e.g., from steel to concrete).

CONCLUSIONS

The paper presents practical procedure for the response prediction and reduction in high-rise buildings under wind loads. To show the applicability of the procedure, aerodynamic loads acting on a quasirectangular high-rise building based on an experimental approach (surface pressure measurement) are used with a mathematical model of the structure for the response prediction and reduction. The building represents a case study of an engineered design of a very slender tower that is instructive. The contributions of this paper can be summarized as follows.(1)The methodology based on HFPI and FEM proposed for the estimation of the response of high-rise buildings under wind loads has the advantage of combining lateral along-wind, lateral cross-wind, and torsional responses altogether. The technique allows for the consideration of any number of modes.(2)Results show that the responses of tall buildings under winds are dominated by the first few modes. Consequently, FEM, 3D lumped mass modeling, and reduced order 3D modeling of tall buildings under wind loads give an accurate assessment of the response provided that the first dominant modes are retained.(3)Results show that the response of tall buildings in the cross-wind direction (lateral response combined simultaneously with torsion) can be higher than the response in the along-wind direction. This reveals the importance of the procedure proposed in this study as many design codes and formula may provide accurate estimate of the along-wind response but less guidance for the estimation of the critical cross-wind and torsional response.(4)The building represents an engineered steel design of a structure that is very much vulnerable to wind loads. This may be due to its low weight as well as high flexibility related to the low dominant frequencies and the high aspect ratio.(5)The building demands TMD with heavier mass and ATMD with higher control force in one lateral direction than the other. This may be attributed to geometry.(6)For the purpose of the use of active control, LQR and fuzzy logic controllers are shown to be effective in enhancing the response reduction over the TMD. ATMDs with fuzzy logic controllers show similar trend like LQR controllers under multidirectional wind loads. In addition, from a design point of view, fuzzy logic controllers do not require the complexity of traditional control systems.(7)The procedure presented in this paper permits the response of tall buildings to be assessed and controlled in the preliminary design stages. This can help decision makers, involved in the design process, to choose among innovative design solutions like structural control, considering several damping techniques, modifying geometry, or even changing materials.

ACKNOWLEDGMENTS

The authors would like to express appreciation to the work team at the Wind Tunnel of Politecnico di Milano, Milan, Italy. The first author wishes to thank Ms Corey Ginsberg, Florida International University, for her helpful comments.

REFERENCES

1.   American Society of Civil Engineers, Minimum Design Loads for Buildings and Other Structures, American Society of Civil Engineers, 2006.

2.   Eurocode 1, "Actions on structures—part 1–4: general actions—wind actions," European Standard prEN 1991-1-4, 2004.

3.   E. Simiu, "Wind loading codification in the Americas: fundamentals for a renewal," in Proceedings of the 11th Americas Conference on Wind Engineering, San Juan, Puerto Rico, USA, June 2009.

4.   S. X. Chen, "A more precise computation of along wind dynamic response analysis for tall buildings built in urban areas," Engineering, vol. 2, pp. 290–298, 2010.

5.   Y. Zhou, T. Kijewski, and A. Kareem, "Aerodynamic loads on tall buildings: interactive database," Journal of Structural Engineering, vol. 129, no. 3, pp. 394–404, 2003.

6.   D. K. Kwon, T. Kijewski-Correa, and A. Kareem, "E-analysis of high-rise buildings subjected to wind loads," Journal of Structural Engineering, vol. 134, no. 7, pp. 1139–1153, 2008.

7.   E. Simiu, R. D. Gabbai, and W. P. Fritz, "Wind-induced tall building response: a time-domain approach,"Wind and Structures, vol. 11, no. 6, pp. 427–440, 2008.

8.   J. R. Wu, Q. S. Li, and A. Y. Tuan, "Wind-induced lateral-torsional coupled responses of tall buildings,"Wind and Structures, vol. 11, no. 2, pp. 153–178, 2008.

9.   M. F. Huang, K. T. Tse, C. M. Chan et al., "An integrated design technique of advanced linear-mode-shape method and serviceability drift

optimization for tall buildings with lateral-torsional modes,"Engineering Structures, vol. 32, no. 8, pp. 2146–2156, 2010.

10. K. T. Tse, P. A. Hitchcock, and K. C. S. Kwok, "Mode shape linearization for HFBB analysis of wind-excited complex tall buildings," Engineering Structures, vol. 31, no. 3, pp. 675–685, 2009.

11. K. M. Lam and A. Li, "Mode shape correction for wind-induced dynamic responses of tall buildings using time-domain computation and wind tunnel tests," Journal of Sound and Vibration, vol. 322, no. 4-5, pp. 740–755, 2009.

12. J. T. P. Yao, "Concept of structural control," American Society of Civil Engineers Journal of the Structural Division, vol. 98, no. 7, pp. 1567–1574, 1972.

13. G. W. Housner, L. A. Bergman, T. K. Caughey et al., "Structural control: past, present, and future,"Journal of Engineering Mechanics, vol. 123, no. 9, pp. 897–971, 1997.

14. Y. Zhou, D. Wang, and X. Deng, "Optimum study on wind-induced vibration control of high-rise buildings with viscous dampers," Wind and Structures, vol. 11, no. 6, pp. 497–512, 2008.

15. A. M. Aly and R. E. Christenson, "On the evaluation of the efficacy of a smart damper: a new equivalent energy-based probabilistic approach," Smart Materials and Structures, vol. 17, no. 4, Article ID 045008, 11 pages, 2008.

16. A. M. Aly, A. Zasso, and F. Resta, "On the dynamics of a very slender building under winds: response reduction using MR dampers with lever mechanism," Structural Design of Tall and Special Buildings, vol. 20, no. 5, pp. 541–553, 2011.

17. R. J. McNamara, "Tuned mass dampers for buildings," American Society of Civil Engineers Journal of the Structural Division, vol. 103, no. 9, pp. 1785–1798, 1977.

18. R. J. Facioni, K. C. S. Kwok, and B. Samali, "Wind tunnel investigation of active vibration control of tall buildings," Journal of Wind Engineering and Industrial Aerodynamics, vol. 54-55, pp. 397–412, 1995.

19. M. Gu and F. Peng, "An experimental study of active control of wind-induced vibration of super-tall buildings," Journal of Wind Engineering and Industrial Aerodynamics, vol. 90, no. 12–15, pp. 1919–1931, 2002.

20. L. T. Lu, W. L. Chiang, J. P. Tang, M. Y. Liu, and C. W. Chen, "Active control for a benchmark building under wind excitations," Journal of Wind Engineering and Industrial Aerodynamics, vol. 91, no. 4, pp. 469–493, 2003.

21. T. T. Soong, Active Structural Control. Theory and Practice, Longman, New York, NY, USA, 1990.

22. J. C. Wu and B. C. Pan, "Wind tunnel verification of actively controlled high-rise building in along-wind motion," Journal of Wind Engineering and Industrial Aerodynamics, vol. 90, no. 12–15, pp. 1933–1950, 2002.

23. A. M. Aly, F. Resta, and A. Zasso, "Active control in a high-rise building under multidirectional wind loads," in Proceedings of the SEI Structures Congress, Vancouver, Canada, 2008.

24. A. M. Aly, On the dynamics of buildings under winds and earthquakes: response prediction and reduction, Ph.D. thesis, Department of Mechanical Engineering, Politecnico di Milano, Milan, Italy, 2009.

25. C. Li, B. Han, J. Zhang, Y. Qu, and J. Li, "Active multiple tuned mass dampers for reduction of undesirable oscillations of structures under wind loads," International Journal of Structural Stability and Dynamics, vol. 9, no. 1, pp. 127–149, 2009.

26. A. Mohtat, A. Yousefi-Koma, and E. Dehghan-Niri, "Active vibration control of seismically excited structures by ATMDS: stability and performance robustness perspective," International Journal of Structural Stability and Dynamics, vol. 10, no. 3, pp. 501–527, 2010.

27. S. J. Park, J. Lee, H. J. Jung, D. D. Jang, and S. D. Kim, "Numerical and experimental investigation of control performance of active mass damper system to high-rise building in use," Wind and Structures, vol. 12, no. 4, pp. 313–332, 2009.

28. S. Homma, J. Maeda, and N. Hanada, "The damping efficiency of vortex-induced vibration by tuned-mass damper of a tower-supported steel stack," Wind and Structures, vol. 12, no. 4, pp. 333–347, 2009.

29. S. Attaway, Matlab: A Practical Introduction to Programming and Problem Solving, Butterworth-Heinemann, Amsterdam, The Netherlands, 2009.

30. G. Diana, S. De Ponte, M. Falco, and A. Zasso, "New large wind tunnel for civil environmental and aeronautical applications," Journal of Wind Engineering and Industrial Aerodynamics, vol. 74–76, pp. 553–565, 1998.

31. A. Zasso, S. Giappino, S. Muggiasca, and L. Rosa, "Optimization of the boundary layer characteristics simulated at Politecnico di Milano boundary layer wind tunnel in a wide scale ratio ranger," inProceedings of the 6th Asia-Pacific Conference on Wind Engineering, Seoul, Korea, September 2005.

32. E. J. Davison, "A method for simplifying linear dynamic systems," IEEE Transactions on Automatic Control, vol. 11, no. 1, pp. 93–101, 1966.

33. J. C. Wu, J. N. Yang, and W. E. Schmitendorf, "Reduced-order H∞ and LQR control for wind-excited tall buildings," Engineering Structures, vol. 20, no. 3, pp. 222–236, 1998.

34. H. T. Nguyen, R. P. Nadipuram, C. L. Walker, and E. A. Walker, A First Course in Fuzzy and Neural Control, Chapman & Hall/CRC, Boca Raton, Fla, USA, 2003.

35. M. Battaini, F. Casciati, L. Faravelli, et al., "Control algorithm and sensor location," in Proceedings of the 2nd World Conference on Structural Control, pp. 1391–1398, Kyoto, Japan, June 1998.

36. B. Samali, M. Al-Dawod, K. C. Kwok, and F. Naghdy, "Active control of cross wind response of 76-story tall building using a fuzzy controller," Journal of Engineering Mechanics, vol. 130, no. 4, pp. 492–498, 2004.

Chapter 12

INTERFACIAL CHARACTERISTICS OF CARBON NANOTUBE-POLYETHYLENE COMPOSITES USING MOLECULAR DYNAMICS SIMULATIONS

Z. Q. Zhang,[1,2] D. K. Ward,[3] Y. Xue,[4] H. W. Zhang,[2] and M. F. Horstemeyer[5]

[1]Center of Micro/Nano Science and Technology, Jiangsu University, Zhenjiang 212013, China

[2]Department of Engineering Mechanics and State Key Laboratory of Structural Analysis for Industrial Equipment, Dalian University of Technology, Liaoning, Dalian 116024, China

[3]Radiation and Nucleation Detection Materials and Analysis Department, Sandia National Laboratories, Livermore, CA 94550, USA

[4]Department of Mechanical and Aerospace Engineering, Utah State University, Logan, UT 84322, USA

[5]Center for Advanced Vehicular Systems, Mississippi State University, Mississippi State, MS 39762, USA

ABSTRACT

The rate-dependent interfacial behavior between a carbon nanotube (CNT) and a polyethylene (PE) matrix is investigated using molecular dynamics (MD) simulations. Various MD simulations were set up to determine the "size" effects on the interfacial properties, such as the molecular weight, or the length of the polymer, the diameter of the CNT, and the simulation model size. The interfacial rate-dependency was probed by applying various relative sliding velocities between the CNT and the polymer. Two quantities, directly obtained from the MD simulations, described the interfacial properties: the critical interfacial shear stress (CISS) and the steady interfacial shear stress (SISS). The simulations show that the SISS was not sensitive to the simulation size. In addition, the CISS was dependent upon the combined factors of the variation in PE stiffness, induced by simulation size changes and the effect of the fixed boundaries of the simulation models. The CISS increases almost linearly with the relative sliding velocity of CNTs. Also, a linear relationship between the SISS and the CNT-sliding velocity is observed when the SISS drops below a critical value. A clear size scaling is observed as the CISS and SISS decrease with increasing CNT radius and increase with the increasing polymer chain length.

INTRODUCTION

Carbon nanotubes (CNTs) have excellent properties, such as very high elastic modulus and tensile strength, which make them suitable for making advanced polymer composites [1]. Recently, several research groups have fabricated and mechanically evaluated CNT-polymer nanocomposites, to explore their potential as advanced lightweight materials [2, 3]. These composites exhibited enhanced moduli, indicating that CNTs carry a portion of the applied load [2–6]. The interfaces between polymer matrix and CNTs transfer load from one to the other and thus play important roles in the nanocomposite's stiffness and strength. At this point, the interfacial properties between CNTs and polymers have not yet been clearly defined for optimization of the composite's mechanical behavior. This paper will examine the CNT/polymer interfacial interactions to determine the dominant mechanisms responsible for load transfer using molecular dynamics (MD) simulations.

The interfacial shear strength, a key property responsible for maximizing the load transfer in short-fiber composites, has been experimentally evaluated using fiber pullout or push-out tests [7, 8]. Experiments have demonstrated that the load transfer mechanism in nanocomposites is weak due to the difficulties in making strong bonds between CNTs and polymer matrices [2, 6, 9, 10]. Thus, CNTs often interact primarily with the polymer matrix through van der Waals forces. The van der Waals forces especially dominate the load transfer capability of the CNT/polymer interface when cross-links between CNTs and polymer matrices are not present. Due to difficulties in devising experiments to study the CNT/polymer interface, molecular mechanics and molecular dynamics simulations have become increasingly popular in investigating the reinforcement mechanisms in nanocomposites [11, 12].

MD simulations offer a method, similar to the CNT pullout from polymer matrix of the previously mentioned experiments, to probe the mechanism of interfacial failure and strengthening. Previous studies, using molecular mechanics simulations of CNT/polymer composite's interfacial binding, predicted that the critical interfacial shear stress (CISS) ranged from 18 to 135 MPa for (10, 10) CNTs embedded in a polymer [13]. The results showed that the binding energy and frictional force played only a minor role in determining the strength of the interface, but the helical polymer conformations in which chains can wrap around nanotubes may enhance nonbonded CNT-polymer interactions. Liao and Li [14] evaluated the interface of a CNT-reinforced polystyrene (PS) composite through molecular mechanics and elasticity calculations. Again, no chemical bonding was considered in the simulation; thus, the interfacial adhesion primarily resulted from van der Waals interaction, mismatch in the coefficients of thermal expansion, and radial deformation induced by atomic interactions. Using MD simulations, Frankland et al. [15] studied the influence of chemical cross-links between a single-walled CNT and a polymer matrix on the interfaces tensile and shear strength. The results suggested that load transfer and modulus of CNT/polymer composites can be effectively enhanced by deliberately increasing the cross-links. Based on this result, they suggested that the chemical bonding between CNTs and polymer matrices is partially responsible for the load transfer at the interface. Frankland and Harik analyzed the entire CNT pullout process by introducing the critical pullout force concept, with the CNT/polymer interfacial sliding described based upon the principle of the Newton's friction law [16]. The results showed that the interfacial interactions during CNT sliding process could be approximated by a linear force-velocity relationship with the slope as the effective viscosity coefficient of the interface. Chowdhury and Okabe [17] also simulated a CNT pullout from a polymer using MD simulations, considering the influence of the polymer matrix density, chemical cross-links at the interface, and geometrical defects in CNTs. They showed that the case without cross-links between the CNT and matrix had an ISS of 1320 MPa which was lower than the case with cross links. Chowdhury and Okabe also showed that the pentagon-heptagon geometrical defect of CNTs has little effect on the interfacial shear stress (ISS). Based on MD simulations, Liu et al. [18] proposed a new boundary element method (BEM) using a cohesive nonbond interface model for CNT/polymer composites. Results demonstrated the usefulness, efficiency, and promise of the developed BEM as a fast numerical tool to characterize the CNT/polymer composites at larger scales. Gou et al. [19] predicted the interfacial bonding of single-walled CNT reinforced epoxy composites using MD simulation based on a cured epoxy resin model. The pullout simulations indicated an effective stress transfer from epoxy resin to the nanotube, which was comparable to their experimental results. Wei [20] studied the temperature-dependent adhesion behavior of CNT/polymer composites using MD simulations and discovered that the ISS resulting from van der Waals interactions increased linearly with the applied tensile strains along the nanotube axis direction and decreased as the temperature increased.

As mentioned above, the interface of CNT/polymer composites has been studied by many research groups, considering the morphology of polymer, the temperature of the simulation systems, the cross-links between the CNTs and polymer, and the defects of the CNTs, using MD simulations, molecular mechanics, and some continuum methods. The contribution of this paper extends the MD simulations on CNT/polymer composites to investigate loading rates and size effects, including the simulation size, the chain length of polymers, and the radius of an embedded CNT.

COMPUTATIONAL METHOD AND MODEL

This work used a classical MD model where the Newtonian equations of motion are solved using the velocity-Verlet algorithm at a time step of 10^{-16} s. The MD simulations were conducted using the software package LAMMPS [21]. The Brenner potential [22] was used for carbon-carbon interactions (or bond) for the CNTs. The modified AMBER potential functions [23] were used to describe the interactions between adjacent PE molecules, with each molecule represented using a united-atom. The total potential energy of the molecular system, E_{tot}, includes bond stretch (bs), bending (be), torsion (to), and the nonbonding potential (van der Waals, vdW) between the united atoms as given by the following:

$$E_{tot} = E_{bs}(r) + E_{be}(\theta) + E_{to}(\phi) + E_{vdW}(\bar{r}),$$

$$E_{bs}(r) = \sum_{atom} \left[k_r (r - r_0) \right],$$

$$E_{be}(\theta) = \sum_{atom} \left[k_\theta (\theta - \theta_0) \right],$$

$$E_{to}(\phi) = \sum_{atom} \left[V_1 \cos\phi + V_2 \cos 2\phi + V_3 \cos 3\phi + V_6 \cos 6\phi \right],$$

$$E_{vdW}(\bar{r}) = 4\varepsilon \left[\left(\frac{\sigma}{r} \right)^{12} - \left(\frac{\sigma}{r} \right)^6 \right],$$

$$(1)$$

where E_{bs} is the two-body potential for bond length r, r0 = 0.1533 nm is the equilibrium bond length, and k_r is the bond stretching stiffness. E_{be} is the three-body potential for bond angle θ, $\theta 0$ = 109.5° is the equilibrium bond angle, and k_θ = 56.25 k_{cal}/mol is the bond bending stiffness. E_{to} is the four-body potential for dihedral angle ϕ. E_{vdW} is another two-body potential for nonbonded distance r, ε is the potential well depth, and σ is distance where the van der Waals energy reaches zero. The van der Waals potential energy $E_{vdW} = E_{vdW} P_E + E_{int}$, where $E_{vdW} P_E$ represents the van der Waals interaction between the PE chains, E_{int} represents the interfacial interaction between the PE chains and the CNT carbon atoms. The potential functions and parameters for the PE are from [17], and, for the nonbonded interfacial interaction between the CNT and the polymer matrix, $\varepsilon = 0.4492$ kJ/mol and $\sigma = 0.401$ nm [24]. A single-walled CNT embedded into an amorphous polyethylene matrix comprised the unit cell of the MD simulation. The CNT spanned the total length of the unit cell, as shown in Figure 1. Periodic boundaries were applied in x and y directions. Two edges of the unit cell in the z direction were fixed to prevent the rigid body motion of the PE, shown as the red boxes in Figure 1. The CNT/PE composites were created according to the following procedure.

(1)   Initially, create the PE amorphous structure; the coordinates of chains united atoms are randomly generated on an FCC lattice, using a random walk method similar to that of Shepherd [25].(2)Obtain the equilibrium configuration of amorphous PE; all the chains are equilibrated for 10^6 MD time steps in the isothermal-isobaric (NPT) ensemble at 300 K to allow for energy relaxation and sufficient entanglement.(3) Generate a cylindrical pore; a cylinder indenter is placed in the center of the simulation box which exerts a force of magnitude $F(r) = -k(r - R)^2$ on each united atom, where k is the specified force constant, r is the distance between the united atoms and the cylinder axis x, $R = R_{CNT} + R_{cutoff}$, R_{CNT} is the radius of the CNT, and $R_{cutoff} = 2.5\ \sigma$ is the van der Waals cutoff distance. (4)Embed an armchair CNT into the cylinder pore; equilibrate the CNT/PE composite for 500,000 MD time steps at 300 K using NPT dynamics to create a zero initial stress state.

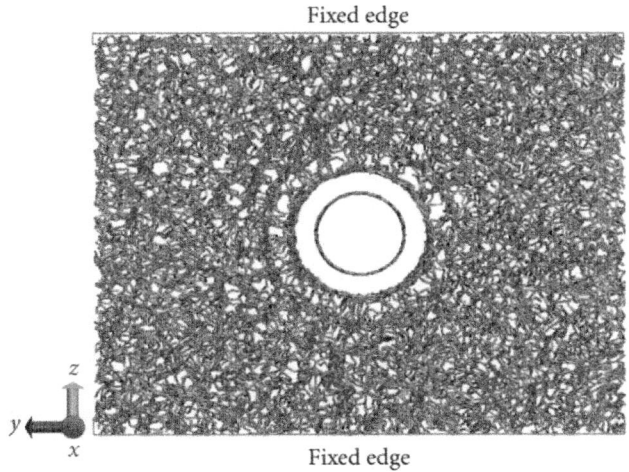

Fixed edge

Fixed edge

Figure 1: Atomistic model of CNT/PE composite. The red boxes show the fixed edges.

Upon reaching an equilibrium configuration for the CNT/PE composites, the temperature of the composite is cooled down from 300 K to ~0 K to avoid any thermal effects of MD simulations during the sliding process [17]. Some small initial stresses, which are dependent on the initial distribution of the PE molecules, may be generated at the interface during this annealing process. After cooling an incremental displacement ranging from 10^{-6} to 10^{-5} nm, which determines the CNT sliding velocity ranging from 10 m/s to 100 m/s, is applied to the CNT atoms to simulate the pullout. In all simulations the CNTs are treated as rigid fibers due to their exceptionally high stiffness compared with PE matrices [18].

SIMULATION RESULTS

Energy Change during the CNT Sliding Process

The ISS is first studied by examining the energetic changes during the sliding of an armchair (10, 10) single-walled CNT (SWCNT) embedded in a block of amorphous PE consisting of 100 chains with 200 monomers/chain (mpc). ISS is calculated using $\tau_I = F_{CNT}/2\pi r L$, where F_{CNT} is the CNT-axial component of the total force on CNT due to the pairwise interactions of PE molecules, r is the CNT radius, and L is the CNT embedded length. In the calculation, the total force on CNT is the sum of the force interaction between the group of carbon atoms of CNT and the group of PE molecules. In MD, forces acting upon atoms are derived from the potential energy which depends on the particles' coordinates. Figure 2 shows ISS as a function of CNT displacement for different sliding velocities. Three distinct stages were observed during the sliding process: (1) a linear increase, which may represent an "elastic" response, before peaking at a possible "yield" or critical stress point, (2) slippage that begins with a transition region where the ISS decreases as the displacement increases, and finally (3) the steady sliding stage where the ISS remains the same as the displacement increases. In the "elastic" regime, the ISS increases almost linearly with displacement until reaching a maximum value. We define this maximum value as the critical interfacial shear stress (CISS). Upon reaching the CISS, the ISS drops sharply suggestive of sliding. After the transition region, the ISS eventually remains constant, which we refer to as a steady sliding stage. During steady sliding stage, ISS oscillates with a period approximately equal to the length of the CNT's unit cell along the CNT axis (2.5 Å), as shown in Figure 2. The results show that the periodic structure of the CNT surface results in an approximate stick-slip sliding [26], with the bonding of the matrix jumping from one unit cell of the CNT to the next. The steady sliding interfacial shear stress (SISS) is obtained by averaging the ISS over the whole steady sliding time steps. The ISS increases as the CNT sliding velocity increases, as shown in Figure 2. Further, the CISS is linearly dependent on the sliding velocity of CNTs (see Figures 4(a), 5(a), 5(c), 5(e), and 6(b)) while SISS nonlinearly increases with the increase of the CNT's sliding velocity (see Figures 4(b), 5(b), 5(d), 5(f), and 6(a)).

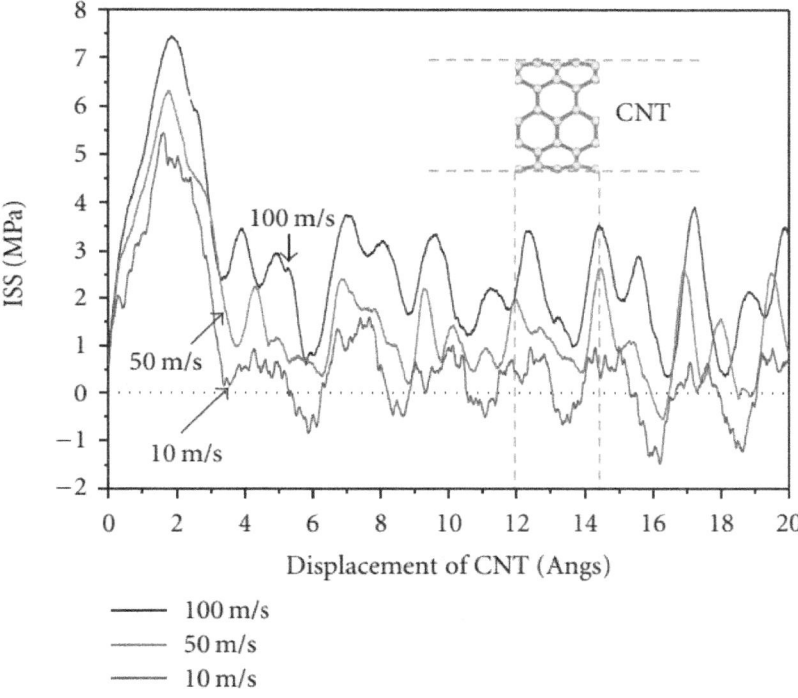

Figure 2: The interfacial shear stress (ISS) as functions of the sliding displacement of (10, 10) SWCNT at different CNT velocities.

Figure 3 shows each component of the potential energy (See the symbols of Figure 3, which are defined in (1) as a function of the CNT sliding displacement with a pulling velocity of 10 m/s. One obvious observation is that only the interfacial energy, E_{int}, varies while tension, bending, torsional, and van der Waals energy do not change upon applying a load. This indicates that the interfacial energy dominates the composite's response, which is described strictly by nonbonded van der Waals interactions between the PE chains and the CNT. The interfacial energy E_{int} increases linearly and reaches a plateaus at the sliding displacement until 3 KCal/mol.

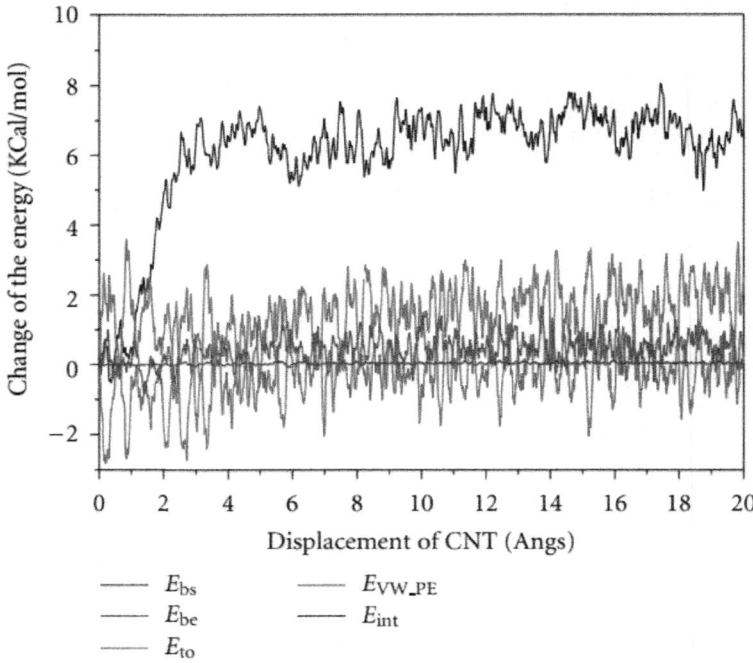

Figure 3: Change of the potential energy as a function of the displacement of SWCNT10, 10 showing that the interfacial energy, E_{int}, changes the most. The sliding velocity of CNT is 10 m/s.

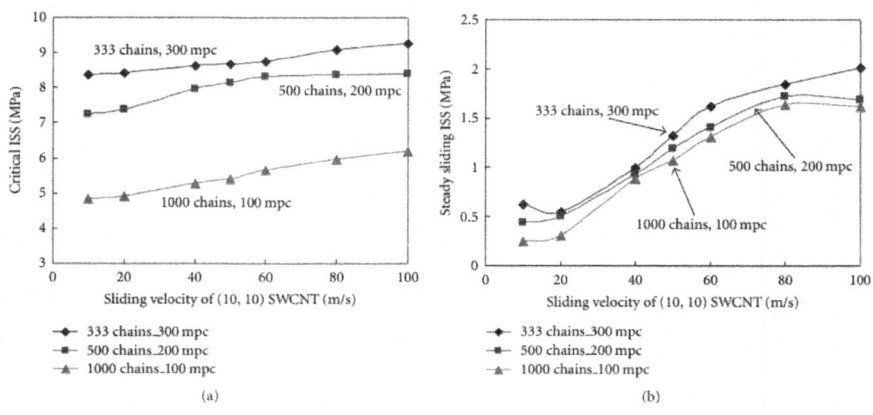

Figure 4: (a) CISS and (b) SISS as functions of sliding velocity of (10, 10) SWCNT at different chain lengths.

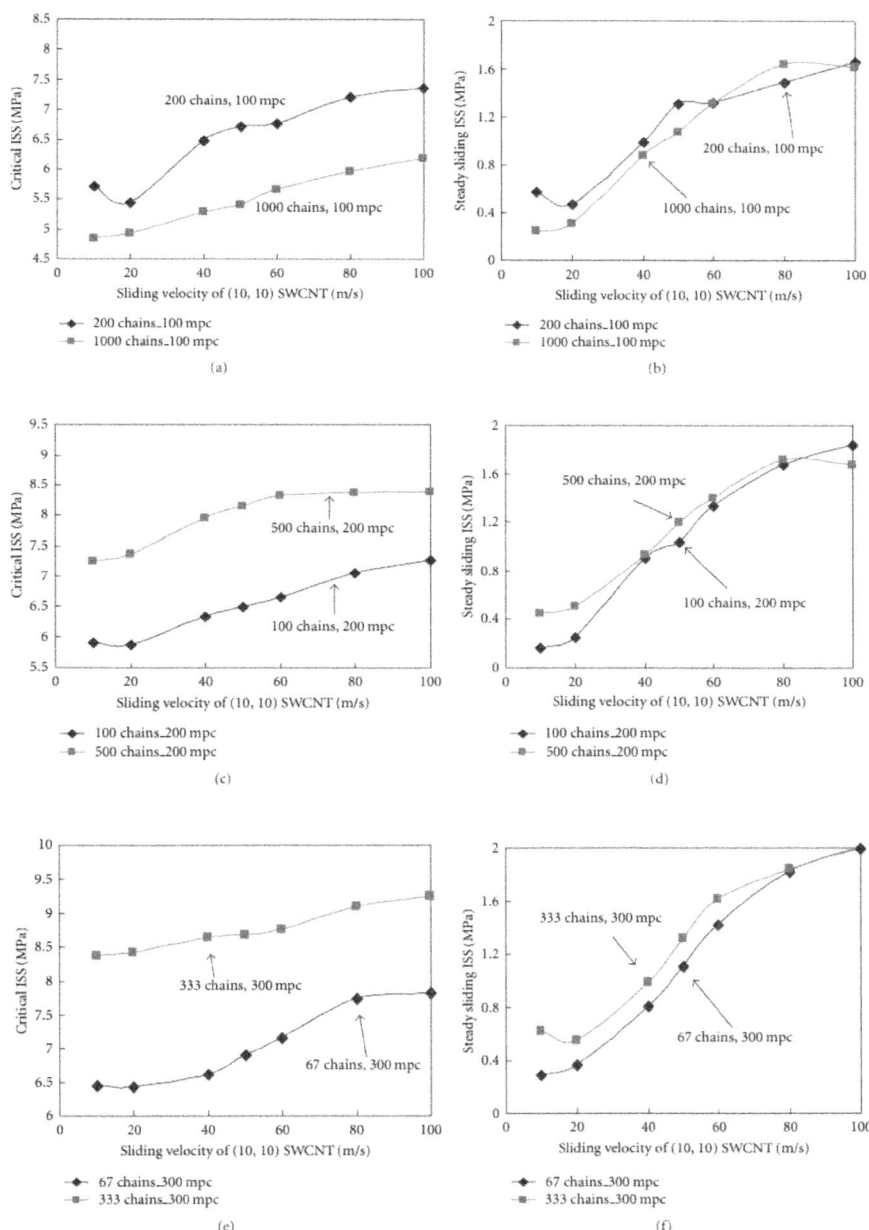

Figure 5: The ISS as a function of the sliding velocity of CNT (10, 10) for the different simulation size.

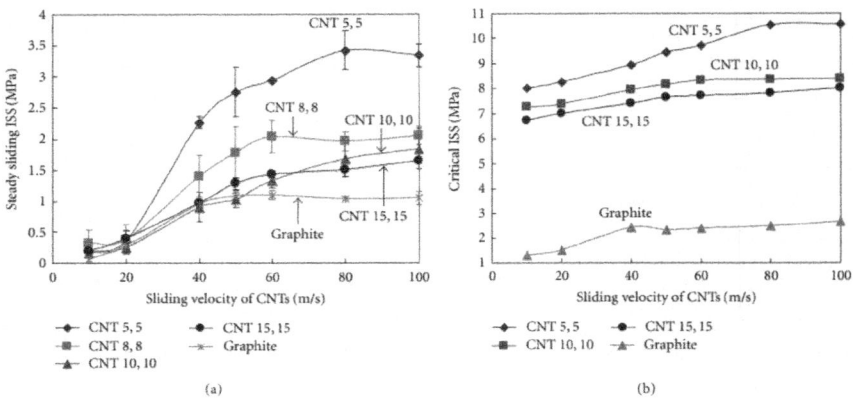

Figure 6: (a) Steady sliding ISS and (b) CISS as functions of the sliding velocity of CNTs at different radii of CNTs, indicated as CNT X, X, where X is the radius in angstroms.

Effect of the PE Chain Length

To investigate the effect of PE chain length on ISS, the PE matrices were built with chain lengths of 100, 200, or 300 molecules/chain (mpc), respectively. An armchair (10, 10) SWCNT was embedded and pulled at velocities ranging from 10 ~ 100 m/s. For each case, in an attempt to remove any effects from the simulation size, the total number of monomers is held constant at 100,000. Also, to ensure the results are not dependent on initial PE velocity profiles, all reported CISS and SISS values are averages of three individual simulations with different initial PE velocity trajectories. Figure 4 shows that both the CISS and SISS increase as the chain length increases. One factor that could increase the CISS is the stiffness of the material. Longer chains tend to lead to a greater number of entanglements in the matrix that could result in a stiffer material. Moreover, the CISS increases linearly with the relative sliding velocity of CNTs, while the relationship between SISS and the CNT-sliding velocity approximately (see Figures 4–6) yields a linear relationship when the SISS is below a critical value. As for the CISS, the shear strain rate of the PE is proportional to the CNT velocity due to the no-slip reinforcement at the PE-CNT interface. In contrast, the SISS is obtained from the steady sliding stage.

Effect of the Simulation Size

The MD simulation cell size effects on the predicted material response are demonstrated from two aspects: the number of monomers and the monomer chain lengths. To capture the effect of the simulation size on the ISS, two simulation sizes, 20,000 and 100,000 monomers, are simulated for chain lengths of 100, 200, and 300 mpc, with the configurations shown in Table 1. The sliding of an armchair (10, 10) SWCNT is examined for velocities ranging from 10 m/s to 100 m/s. Figure 5 shows the CISS and SISS for all cases as a function of the CNT sliding velocity. For the cases of 200 and 300 mpc, enlarging the simulation size from 100 chains to 500 chains clearly increases the CISS while the SISS changes only slightly (see Figures 5(c) and 5(d)). In contrast, for the simulations of chains with 100 mpc, the CISS decreased when the number of chains increased from 200 to 1000 (see Figure 5(a)). In short, for increasing simulation size, the CISS decreases for shorter chain lengths but increases for longer chains. In contrast, the SISS is not sensitive to the simulation size for these configurations (see Figures 5(b), 5(d), and 5(f)).

Table 1: Simulation sizes for different configurations. l_x, l_y and l_z are the length of the simulation box in x, y and z directions

Simulation samples	Simulation box size = $l_x^* l_y^* l_z$ (Å³)
PE: 200 chains, 100 mpc, CNT (10, 10)	60.8 × 76.3 × 120.4
PE: 1000 chains, 100 mpc, CNT (10, 10)	144.4 × 135.6 × 140.8
PE: 100 chains, 200 mpc, CNT (10, 10)	96.3 × 68.1 × 85.3
PE: 500 chains, 200 mpc, CNT (10, 10)	134.2 × 142.2 × 142.9
PE: 67 chains, 300 mpc, CNT (10, 10)	81.0 × 80.8 × 84.9
PE: 333 chains, 300 mpc, CNT (10, 10)	134.3 × 139.6 × 144.7

Effect of the CNT Radius

Armchair (5, 5), (8, 8), (10, 10), and (15, 15) SWCNTs, with corresponding radii of 0.346, 0.554, 0.692, and 1.038 nm, respectively, were simulated to study the role of CNT radii on interfacial properties. Figure 6 shows that the SISS, as a function of the sliding velocity for different radii of CNTs, decreased as the CNT radius increased. For higher sliding velocities, the ISS plateaus and the differences in ISS due to CNT radius are evident. Interestingly, in the high sliding velocity region, the distance between the curves of (5, 5) and (8, 8) CNTs is larger than that between the curves of (8, 8) and (10, 10) CNTs, while there is only a little difference between the curves of (10, 10) and (15, 15) CNTs. It implies that for an increasing CNT radius, there is a lower limit for the SISS for a curvature approaching zero (a flat graphite sheet). To show the lower limit the ISS is calculated for a graphite sheet embedded in a PE matrix and displaced at the same range of constant velocities as above. The results again show that the SISS increases as the sliding velocity increases. It indicates that the SISS in the case of graphite sheet is a lower bound value (1.1 MPa) in the high sliding velocity region (see Figure 6(a)). Furthermore, the effect of the radius of CNTs on CISS is investigated, as shown in Figure 6(b). The CISS also decreases as the CNT radius increases, and the lower bound value of CISS (2.4 MPa) is obtained from the graphite sheet sliding process.

DISCUSSIONS

The interfacial behavior of CNT-polymer composites, such as the SISS and CISS, is affected significantly by the MD simulation model. The fundamental influential factors for the interfacial behavior are studied in the following.

Effect of Polymer Chain Length

Supplemental tensile simulations of bulk polymer with chain lengths of 100, 200, and 300 mpc have been performed to determine the matrix elastic properties for a loading strain rate of 10^8/s. The results show that the stiffness of the PE increases significantly as the chain length increases from 100 to 200 mpc while there is a little change when the chain length increases from 200 to 300 mpc, as shown in Figure 7. This is in agreement with previous work for similar MD simulations of PE [27].

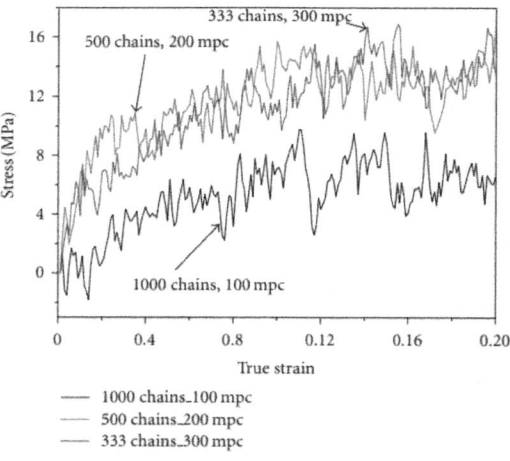

Figure 7: The comparison between the stress-strain behaviors of PE systems consisting of 1000 chains each with 100 monomers, 500 chains each with 200 monomers, and 333 chains each with 300 monomers. The strain loading rate is 10^8/s.

The increases in stiffness parallel the results of chain length effects on the CISS, which increases with increasing chain length from 100 to 200 mpc while there is a little increase from 200 to 300 mpc (see Figure 4). This suggests that the increase of the ISS with increasing chain length is a result of the increasing PE stiffness.

Effect of Polymer Molecular Density at the Interface

One method for explaining the increase in shear stress is to examine the proximity of PE molecules to the surface of the CNT, which can be described using a radial distribution function (RDF) (r). From the MD simulations, the expression for evaluating (r) from the output data is given as follows:

$$g\left(r\right) = \frac{\sum_{i=1}^{N} N_i\left(r, \Delta r\right)}{N\rho V\left(r, \Delta r\right)},$$

(2)

where N is the total number of atoms in the simulation system, $\rho=N/V$ the atom number density with V being the representative volume, N_i the number of atoms found in a spherical shell of radius r and thickness Δr, with the cell centered on atom I, and $V(r,\Delta r)$ the volume of the spherical shell. In Figure 8, the slight increase in the RDF indicates the presence of more PE atoms at the interface for the critical point of CNT slippage relative to the relaxed state. This increased density at the critical point results in a higher interfacial energy thus, facilitating a larger critical ISS than the sliding ISS (see Figure 2) during the CNT-pulling process.

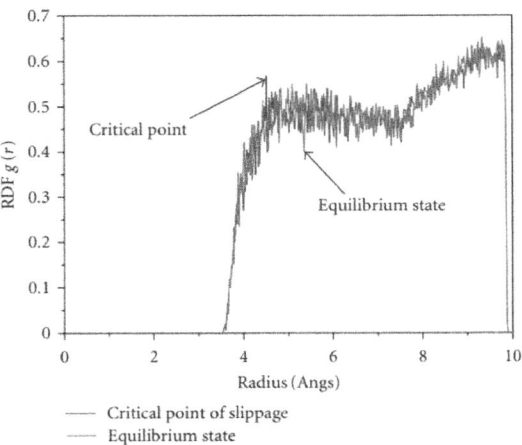

Figure 8: The radial distribution function (RDF) in the case of armchair (10, 10) CNT sliding in the PE with 100 chains, 200 monomers per chain. The sliding velocity of CNT is 10 m/s.

Effect of the Simulation Model Sizes

There are two factors influencing the CISS as the simulation size changes. One factor is the stiffness of PE and the other is the fixed boundaries used to prevent the PE chains from moving with CNTs in our simulation model (see Figure 1). A series of tension tests are again carried out at a strain rate of 10^8/s to investigate the effect of the simulation size on the stiffness of PE. A set of stress-strain curves is obtained for each chain length and simulation size and shown in Figure 9. For small chain length, 100 mpc, enlarging the simulation size has almost no effect on the elastic modulus of PE (see Figure 9(a)), while there is a prominent increase for the chain length of >200 mpc (see Figures 9(b) and 9(c)).

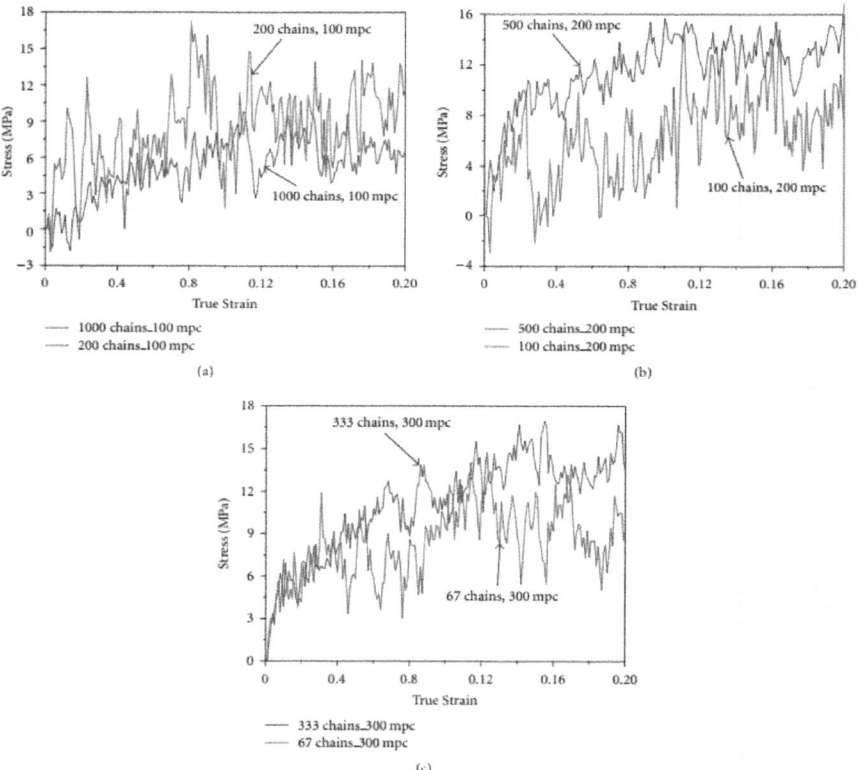

Figure 9: Comparisons between the stress-strain behaviors of PE systems consisting of (a) 200 and 1000 chains each with 100 monomers, (b) 100 and 500 chains each with 200 monomers, and (c) 67 and 333 chains each with 300 monomers. The strain loading rate is 10^8/s.

The second effect due to the fixed boundary is caused by a localization of the polymer deformation near the nanotube surface. The displacements of PE molecules along the CNT sliding direction versus the distance from the tube wall to the fixed boundary at the point of CISS are plotted in Figure 10. For chains length of 100 mpc (See Figure 10(a)), the displacement of the PE molecules at the interface is smaller for the simulation size. This decrease in displacement implies that the fix boundary effect dominants for the shorter chain length cases due to the increase of the distance from the fixed boundary to the tube wall. Oppositely, for the chain length of 300 mpc (see Figure 10(c)), the displacement of the interfacial PE molecules increases with increasing the number of chains from 67 to 333, which reveals that the increase of the stiffness of PE dominates for the longer chain length cases, resulting in an increase in CISS with increasing simulation size. The chain length of 200 mpc (see Figure 10(b)) is similar to the 300 mpc case.

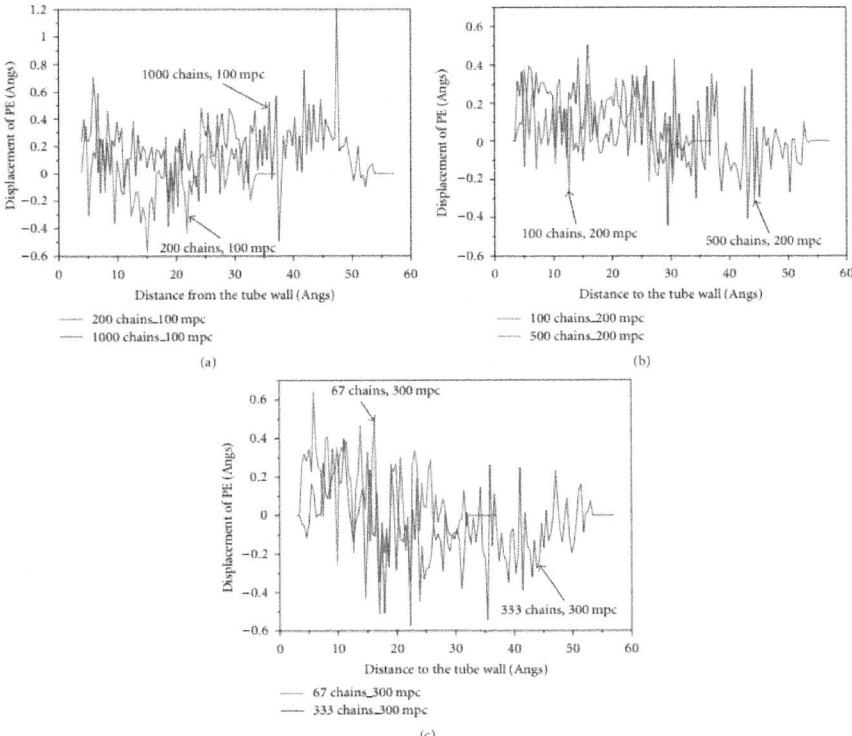

Figure 10: Comparisons between the displacements of PE molecules each with (a) 100 monomers, (b) 200 monomers, and (c) 300 monomers in the CNT sliding direction at different simulation sizes.

Effect of CNT Radii

To study the various mechanisms of the SISS with an increasing CNT radius, we trace back to the calculation of ISS, $\tau_{ISS}=F_{CNT}/A_{int}$, where F_{CNT} is the total force upon a CNT A_{int} is the interface area. For one carbon atom of a CNT, the occupied area is a constant value, which is independent of the CNT radius. Thus, the ISS is directly proportional to the force upon a carbon atom of CNT. Moreover, the force upon a carbon atom is directly proportional to the interaction energy, when it experiences the same displacement. Thus, the ISS is directly proportional to the interaction energy for one carbon atom of CNT. As shown schematically in Figure11, the PE region of interaction with a single appointed carbon atom decreases with enlarging the radius of CNTs. Hence, the ISS decreases as the CNT radius increases, or the ratio of volume of interaction versus CNT surface area decreases as the CNT radius increases. The reason the graphite sheet sets the lower limit for the ISS is that the PE region of interaction for a single carbon atom is smaller than any other radius.

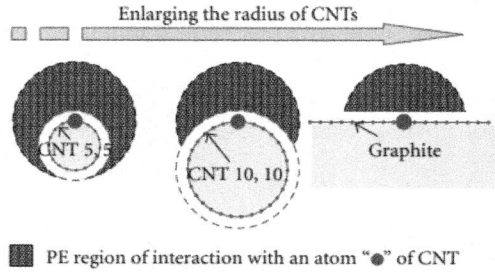

Figure 11: The schematic of PE regions of interaction with a carbon atom of CNT with enlarging the radius of CNTs. The graphite sheet can be seen as a CNT with infinitely large radius.

SUMMARY

MD simulations are used to study the interfacial characteristics of CNT/PE composites using a nonbonded interfacial model. The interfacial behaviors between a CNT and a PE matrix are described as a function of the MD simulation model size, sliding velocity of CNTs, radius of CNTs, and polymer chain length. The CISS exhibits a linear dependence with the sliding velocity of the CNTs. The SISS is not sensitive to the simulation size while the CISS is significantly affected by the simulation size, resulting from a trade-off between the proximity of the rigid boundary and the stiffness of PE. Both the CISS and SISS decrease as the CNT radius increases but increase as the polymer chain length increases.

ACKNOWLEDGMENT

This paper was conducted collaboratively when the first three authors worked at the Center for Advanced Vehicular Systems (CAVS), Mississippi State University and thus would like to thank CAVS for the support.

REFERENCES

1.   K. T. Lau and D. Hui, "The revolutionary creation of new advanced materials—carbon nanotube composites," Composites Part B, vol. 33, no. 4, pp. 263–277, 2002.

2.   L. S. Schadler, S. C. Giannaris, and P. M. Ajayan, "Load transfer in carbon nanotube epoxy composites,"Applied Physics Letters, vol. 73, no. 26, pp. 3842–3844, 1998.

3.   M. S. P. Shaffer and A. H. Windle, "Fabrication and characterization of carbon nanotube/poly(vinyl alcohol) composites," Advanced Materials, vol. 11, no. 11, pp. 937–941, 1999.

4.   X. Gong, J. Liu, S. Baskaran, R. D. Voise, and J. S. Young, "Surfactant-assisted processing of carbon nanotube/polymer composites," Chemistry of Materials, vol. 12, no. 4, pp. 1049–1052, 2000.

5.   R. Haggenmueller, H. H. Gommans, A. G. Rinzler, J. E. Fischer, and K. I. Winey, "Aligned single-wall carbon nanotubes in composites by melt processing methods," Chemical Physics Letters, vol. 330, no. 3-4, pp. 219–225, 2000.

6.   D. Qian, E. C. Dickey, R. Andrews, and T. Rantell, "Load transfer and deformation mechanisms in carbon nanotube-polystyrene composites," Applied Physics Letters, vol. 76, no. 20, pp. 2868–2870, 2000. ·

7.   M. R. Piggott, "A new model for interface failure in fibre-reinforced polymers," Composites Science and Technology, vol. 55, no. 3, pp. 269–276, 1995.

8.   V. T. Bechel and N. R. Sottos, "A comparison of calculated and measured debond lengths from fiber push-out tests," Composites Science and Technology, vol. 58, no. 11, pp. 1727–1739, 1998.

9.   E. T. Thostenson, Z. Ren, and T. W. Chou, "Advances in the science and technology of carbon nanotubes and their composites: a review," Composites Science and Technology, vol. 61, no. 13, pp. 1899–1912, 2001. ·

10. P. M. Ajayan, L. S. Schadler, C. Giannaris, and A. Rubio, "Single-walled carbon nanotube-polymer composites: strength and weakness," Advanced Materials, vol. 12, no. 10, pp. 750–753, 2000.

11. Q. Zheng, Q. Xue, K. Yan, X. Gao, Q. Li, and L. Hao, "Effect of chemisorption on the interfacial bonding characteristics of carbon nanotube-polymer composites," Polymer, vol. 49, no. 3, pp. 800–808, 2008.

12. M. Al-Haik, M. Y. Hussaini, and H. Garmestani, "Adhesion energy in carbon nanotube-polyethylene composite: effect of chirality," Journal of Applied Physics, vol. 97, no. 7, Article ID 074306, 5 pages, 2005. ·

13. V. Lordi and N. Yao, "Molecular mechanics of binding in carbon-nanotubepolymer composites," Journal of Materials Research, vol. 15, no. 12, pp. 2770–2779, 2000.

14. K. Liao and S. Li, "Interfacial characteristics of a carbon nanotube-polystyrene composite system," Applied Physics Letters, vol. 79, no. 25, pp. 4225–4227, 2001.

15. S. J. V. Frankland, A. Caglar, D. W. Brenner, and M. Griebel, "Molecular simulation of the influence of chemical cross-links on the shear strength of carbon nanotube-polymer interfaces," Journal of Physical Chemistry B, vol. 106, no. 12, pp. 3046–3048, 2002.

16. S. J. V. Frankland and V. M. Harik, "Analysis of carbon nanotube pull-out from a polymer matrix," Surface Science, vol. 525, no. 1-3, pp. L103–L108, 2003.

17. S. C. Chowdhury and T. Okabe, "Computer simulation of carbon nanotube pull-out from polymer by the molecular dynamics method," Composites Part A, vol. 38, no. 3, pp. 747–754, 2007.

18. Y. J. Liu, N. Nishimura, D. Qian, N. Adachi, Y. Otani, and V. Mokashi, "A boundary element method for the analysis of CNT/polymer composites with a cohesive interface model based on molecular dynamics," Engineering Analysis with Boundary Elements, vol. 32, no. 4, pp. 299–308, 2008.

19. J. Gou, B. Minaie, B. Wang, Z. Liang, and C. Zhang, "Computational and experimental study of interfacial bonding of single-walled nanotube reinforced composites," Computational Materials Science, vol. 31, no. 3-4, pp. 225–236, 2004.

20. Wei, "Adhesion and reinforcement in carbon nanotube polymer composite," Applied Physics Letters, vol. 88, no. 9, Article ID 093108, 2006.

21. S. Plimpton, "Fast parallel algorithms for short-range molecular dynamics," Journal of Computational Physics, vol. 117, no. 1, pp. 1–19, 1995.

22. K. Liao and S. Li, "Interfacial characteristics of a carbon nanotube-polystyrene composite system,"Applied Physics Letters, vol. 79, no. 25, pp. 4225–4227, 2001.

23. S. Kuwajima, H. Noma, and T. Ohsaka, in Proceedings of the 4th Symposium of the Society of Computer Chemistry, pp. 53–56, Japan, 1994.

24. M. P. Allen and D. J. Tildesley, Computer Simulation of Liquids, Clarendon Press, 1987.

25. J. E. Shepherd, Multiscale modeling of the deformation of semi-crystalline polymers, dissertation, Georgia Institute of Technology, 2006.

26. Gourdon and J. N. Israelachvili, "Transitions between smooth and complex stick-slip sliding of surfaces," Physical Review E, vol. 68, no. 2, Article ID 021602, 10 pages, 2003.

27. Hossain, M. A. Tschopp, D. K. Ward, J. L. Bouvard, P. Wang, and M. F. Horstemeyer, "Molecular dynamics simulations of deformation mechanisms of amorphous polyethylene," Polymer, vol. 51, no. 25, pp. 6071–6083, 2010.

Chapter 13

UNMANNED AERIAL VEHICLE NAVIGATION USING WIDE-FIELD OPTICAL FLOW AND INERTIAL SENSORS

Matthew B. Rhudy,[1] Yu Gu,[2] Haiyang Chao,[3] and Jason N. Gross[4]

[1]Division of Engineering, Pennsylvania State University, Reading, PA 19610, USA

[2]Department of Mechanical and Aerospace Engineering and Lane Department of Computer Science and Electrical Engineering at WVU, Morgantown, WV 26506, USA

[3]Aerospace Engineering Department, University of Kansas, Lawrence, KS 66045, USA

[4]Department of Mechanical and Aerospace Engineering at West Virginia University, Morgantown, WV 26506, USA

ABSTRACT

This paper offers a set of novel navigation techniques that rely on the use of inertial sensors and wide-field optical flow information. The aircraft ground velocity and attitude states are estimated with an Unscented Information Filter (UIF) and are evaluated with respect to two sets of experimental flight data collected from an Unmanned Aerial Vehicle (UAV). Two different formulations are proposed, a full state formulation including velocity and attitude and a simplified formulation which assumes that the lateral and vertical velocity of the aircraft are negligible. An additional state is also considered within each formulation to recover the image distance which can be measured using a laser rangefinder. The results demonstrate that the full state formulation is able to estimate the aircraft ground velocity to within 1.3 m/s of a GPS receiver solution used as reference "truth" and regulate attitude angles within 1.4 degrees standard deviation of error for both sets of flight data.

INTRODUCTION

Information about the velocity and attitude of an aircraft is important for purposes such as remote sensing [1], navigation, and control [2]. Traditional low-cost aircraft navigation relies on the use of both inertial sensors and Global Positioning System (GPS) [3–5]. While GPS can provide useful information to an aircraft system, this information is not always available or reliable in certain situations, such as flying in urban environments or other GPS-denied areas (e.g., under radio-frequency jamming or strong solar storm). GPS is not self-contained within the aircraft system; rather the information comes from external satellites. Insects, such as the honeybee, have demonstrated impressive capabilities in flight navigation without receiving external communications [6]. One significant information source that is used by insects as well as birds is vision [6–8]. This information can also be made available to an aircraft through the use of onboard video cameras. The challenge with this information rich data is correctly processing and integrating the vision data with the other onboard sensor measurements [9].

Vision data can be processed using feature detection algorithms such as the Scale-Invariant Feature Transform (SIFT) [10] to obtain optical flow vectors, as well as other techniques. Optical flow is useful for aircraft systems because it is rich in navigation information, simple to represent, and easy to compute [11]. One of the benefits of this information is that it can be used in order to extract velocity information about the aircraft, which in turn can be used for aircraft positioning. This optical flow information has been used for autonomous navigation applications such as relative heading and lateral position estimation of a quadrotor helicopter [12,13]. Another work has considered the use of optical flow for UAV take-off and landing [14] and landmark navigation [15]. Another potential benefit of optical flow is that it implicitly contains information about the aircraft attitude angles. This implicit information has been used in related work for UAV attitude estimation using horizon detection and optical flow along the horizon line [16, 17] and pose estimation for a hexacopter [18], a lunar rover [19], and spacecraft [20]. While this work is useful, these vehicles contain significantly different dynamic characteristics than a typical airplane. Due to this, more analysis of the application of optical flow for airplane applications is necessary.

This work presents a combined velocity and attitude estimation algorithm using wide-field optical flow for airplanes that does not require horizon detection, which is useful because the horizon does not need to be visible in the image frame in order to obtain attitude information. The algorithm relies on the optical flow computed using a downward facing video camera, measurements from a laser range finder and an Inertial Measurement Unit (IMU) that are mounted in parallel to the camera axis, and a flat ground assumption to determine information about the aircraft velocity and attitude. Many of the existing experiments for optical flow and inertial sensor fusion are done using helicopter platforms and focus on position and velocity estimation [21, 22]. This work considers an airplane system rather than a helicopter, which contains a significantly different flight envelope and dynamics. Additionally, the regulation of attitude information through the use of optical flow is considered, which is not typically done in existing applications. This work takes advantage of all detected optical flow points in the image plane, including wide-field optical flow points which were often omitted in previous works [23–25]. These wide-field optical flow points are of significant importance for attitude estimation, since they contain roll and pitch information that is not observable from the image center. Although this work considers the use of a laser range finder to recover the distance between the image scene and the camera, it is possible to determine this information using other techniques [26]. In fact, it has been demonstrated that the scale is an observable mode for the vision and IMU data fusion problem [27]. The presented formulation was originally offered in its early stages of development in [28]. Since this original publication, the implementation and tuning of the formulation have been refined, and additional results have been generated. In particular, a simplified formulation is offered which reduces the filter states, and the inclusion of a range state is considered. The main contribution of this paper is the analysis of a stable vision-aided solution for the velocity and attitude determination without the use of GPS. This solution is verified with respect to two sets of actual UAV flight testing data.

The rest of this paper is organized as follows. Section 2 presents the different considered formulations and framework for this problem. Section 3 describes the experimental setup which was used to collect data for this study. The results are offered in Section 4 followed by a conclusion in Section 5.

PROBLEM FORMULATION

Optical Flow Equations

Optical flow is the projection of 3D relative motion into a 2D image plane. Using the pinhole camera model, the 3D position (η_x, η_y, η_z) in the 3D camera body frame can be mapped into the 2D image plane with coordinates (μ, ν) using

$$\mu = f\frac{\eta_x}{\eta_z},$$

$$\nu = f\frac{\eta_y}{\eta_z},$$

(1)

where μ,], and f are given in pixels and f is the focal length. For a downward looking camera that is parallel to the aircraft z-axis, and with a level and flat ground assumption, the optical flow equations have been derived [29]:

$$
\begin{bmatrix} \dot{\mu} \\ \dot{v} \end{bmatrix} = \frac{f + v \tan\phi - \mu \tan\theta / \cos\phi}{\eta_z} \begin{bmatrix} -u + \dfrac{\mu w}{f} \\ -v + \dfrac{vw}{f} \end{bmatrix}
$$

$$
- \begin{bmatrix} fq - rv - \dfrac{p\mu v}{f} + \dfrac{q\mu^2}{f} \\ -fp + r\mu - \dfrac{pv^2}{f} + \dfrac{q\mu v}{f} \end{bmatrix},
$$

(2)

where ϕ, θ are the roll and pitch angles, p, q, r are the roll, pitch, and yaw body-axis angular rates, u, V, w, are the bodyaxis ground velocity components of the aircraft, and μ,] are the components of optical flow in the 2D image plane, given in pixels/sec. This equation captures the relationship between optical flow at various parts of the image plane with other pieces of navigation information. By considering only the area close to the image center ($\mu \approx 0$,] ≈ 0), the narrow-field optical flow model can be simplified [23–25]; however, this removes the roll and pitch dependence of the equation and is therefore not desirable for attitude estimation purposes.

State Space Formulation and Stochastic Modeling

This work considers the simultaneous estimation of body-axis ground velocity components (u, V, w) and Euler attitude angles (ϕ, θ, ψ). This estimation is performed through the fusion of Inertial Measurement Unit (IMU) measurements of body-axis accelerations (a_x, a_y, a_z) and angular rates (p, q, r), laser rangefinder range measurements (L), and n sets of optical flow measurements (μ,])i , where $i = 1, 2, \ldots, n$. The value of n varies with each time step based on how many features in the frame can be used for optical flow calculation. Using these values, the state space model of the system is formulated with the following state vector, x, bias state vector, b, input vector, u, optical flow input vectors, di, and output vectors, zi:

$$
\mathbf{x} = \begin{bmatrix} u & v & w & \phi & \theta & \psi & \mathbf{b}^T \end{bmatrix}^T,
$$

$$
\mathbf{b} = \begin{bmatrix} b_{a_x} & b_{a_y} & b_{a_z} & b_p & b_q & b_r & b_L \end{bmatrix}^T,
$$

$$
\mathbf{u} = \begin{bmatrix} a_x & a_y & a_z & p & q & r & L \end{bmatrix}^T,
$$

$$
\mathbf{d}_i = \begin{bmatrix} \mu & v \end{bmatrix}_i^T,
$$

$$
\mathbf{z}_i = \begin{bmatrix} \dot{\mu} & \dot{v} \end{bmatrix}_i^T.
$$

(3)

A diagram describing the definition of the range coordinate, , is provided in Figure 1. Note that the range coordinate, L, is equivalent to the camera z coordinate, η_z.

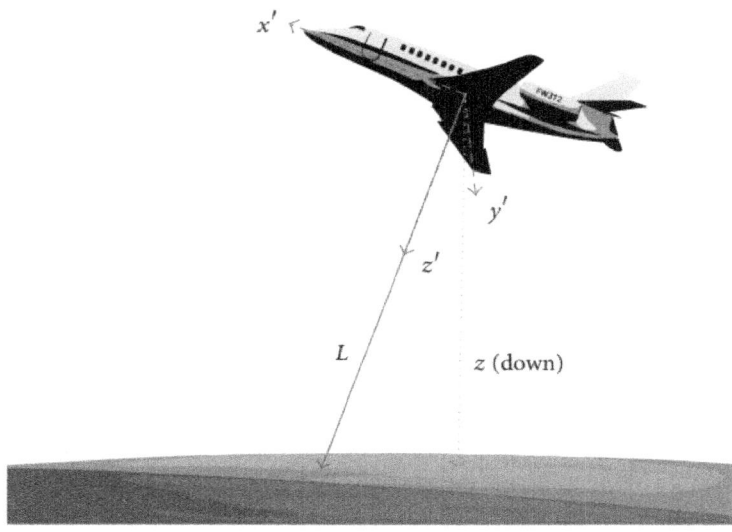

$\leftarrow - -$ Body coordinates
$\leftarrow \cdots$ NED coordinates
$\leftarrow\!\!\!-\!\!-$ Range coordinate

Figure 1: Diagram of the range coordinate.

In order to determine the dynamics of the velocity states, the time derivative of the velocity vector observed from the fixed navigation frame is equal to the time rate of change as observed from the moving body axis frame plus the change caused by rotation of the frame [30]:

$$\frac{d}{dt}\left(\begin{bmatrix} u \\ v \\ w \end{bmatrix}\right) = \begin{bmatrix} \dot{u} \\ \dot{v} \\ \dot{w} \end{bmatrix} + \begin{bmatrix} p \\ q \\ r \end{bmatrix} \times \begin{bmatrix} u \\ v \\ w \end{bmatrix}.$$

$$(4)$$

The IMU measures the acceleration with respect to the fixed gravity vector, as inwhere is the rotation matrix from the navigation frame to the body frame:

$$\begin{bmatrix} a_x \\ a_y \\ a_z \end{bmatrix} = \frac{d}{dt}\left(\begin{bmatrix} u \\ v \\ w \end{bmatrix}\right) + C_n^b \begin{bmatrix} 0 \\ 0 \\ -g \end{bmatrix},$$

$$(5)$$

where C_n^b is the rotation matrix from the navigation frame to the body frame:

$$C_n^b = \begin{bmatrix} \cos\theta\cos\psi & \cos\theta\sin\psi & -\sin\theta \\ -\cos\phi\sin\psi + \sin\phi\sin\theta\cos\psi & \cos\phi\cos\psi + \sin\phi\sin\theta\sin\psi & \sin\phi\cos\theta \\ \sin\phi\sin\psi + \cos\phi\sin\theta\cos\psi & -\sin\phi\cos\psi + \cos\phi\sin\theta\sin\psi & \cos\phi\cos\theta \end{bmatrix}. \tag{6}$$

Combining these results gives the dynamics for the velocity states [31]:

$$\begin{bmatrix} \dot{u} \\ \dot{v} \\ \dot{w} \end{bmatrix} = \begin{bmatrix} rv - qw + a_x - g\sin\theta \\ pw - ru + a_y + g\sin\phi\cos\theta \\ qu - pv + a_z + g\cos\phi\cos\theta \end{bmatrix}. \tag{7}$$

The dynamics of the attitude states are defined using [32]

$$\begin{bmatrix} \dot{\phi} \\ \dot{\theta} \\ \dot{\psi} \end{bmatrix} = \begin{bmatrix} 1 & \sin\phi\tan\theta & \cos\phi\tan\theta \\ 0 & \cos\phi & -\sin\phi \\ 0 & \sin\phi\sec\theta & \cos\phi\sec\theta \end{bmatrix} \begin{bmatrix} p \\ q \\ r \end{bmatrix}. \tag{8}$$

To define the dynamics for the bias parameters, a first-order Gauss-Markov noise model was used. In a related work [33], the Allan deviation [34] approach presented in [35, 36] was used to determine the parameters of the first-order Gauss-Markov noise model for the dynamics of the bias on each IMU channel. The Gauss-Markov noise model for each sensor measurement involves two parameters: a time constant and a variance of the wide-band sensor noise. Using this model, the dynamics for the bias parameters are given by

$$\mathbf{b}_k = \mathbf{b}_{k-1} e^{-T_s/\tau} + \mathbf{n}_{k-1}, \tag{9}$$

where τ is a vector of time constants and n is a zero-mean noise vector with variance given by a diagonal matrix of the variance terms for each sensor. The time constant and variance terms were calculated in [33] for each channel of the same IMU that was considered for this study. The state dynamic equations have been defined in continuous-time using the following format:

$$\dot{\mathbf{x}} = \mathbf{f}_c(\mathbf{x}, \mathbf{u}), \tag{10}$$

where $\mathbf{f}_c$ is the nonlinear continuous-time state transition function. In order to implement these equations in a discrete-time filter, a first-order discretization is used [37]:

$$\mathbf{x}_k = \mathbf{x}_{k-1} + T_s \mathbf{f}_c(\mathbf{x}_{k-1}, \mathbf{u}_{k-1}) \triangleq \mathbf{f}(\mathbf{x}_{k-1}, \mathbf{u}_{k-1}), \tag{11}$$

where k is the discrete time index, f is the nonlinear discretetime state transition function, and T_s is the sampling time of the system.

To formulate the observation equations, optical flow information is utilized. In particular, each optical flow point identified from vision data consists of four values: (μ, ν). These values are obtained using a point matching method [38] and the Scale-Invariant Feature Transform (SIFT) algorithm [10]. Note that the method for optical flow generation is not the emphasis of this research [38]; therefore, any other optical flow algorithm can be used similarly within the proposed estimator, without any loss of generality.

During the state estimation process, the image plane coordinates (μ, ν) are taken as inputs to the observation equation, allowing the optical flow $(\dot{\mu}, \dot{\nu})$ to be predicted at that point in the image plane using (2), where η_z is provided by the laser rangefinder measurement, L. These computed observables are then compared with the optical flow measurements of $(\dot{\mu}, \dot{\nu})$ from the video in order to determine how to update the states. Since multiple optical flow points can be identified within a single time step, this creates a set of n_k observation equations, where n_k is the number of optical flow points at time step k. Since (7) and (8) are derived from kinematics, the only uncertainty that must be modeled is due to the input measurements. Therefore, the input vector is given by

$$\mathbf{u}_{k-1} = \hat{\mathbf{u}}_{k-1} + \mathbf{b}_{k-1}, \tag{12}$$

where $\hat{\mathbf{u}}$ is the measured input vector and b is the vector of sensor biases which follow a first order Gauss-Markov noise model as determined in [33].

The uncertainty in the measurements is due to the errors in the optical flow estimation from the video. It is assumed that each optical flow measurement $\mathbf{y}_i$ has an additive measurement noise vector, $\mathbf{v}_i$, with corresponding covariance matrix, $\mathbf{R}_i$. For this study, it is also assumed that each optical flow measurement carries equal uncertainty and that errors along the two component directions of the image plane also have equal uncertainty and are uncorrelated; that is,

$$\mathbf{R}_i = \mathbf{R} = R\mathbf{I}, \tag{13}$$

where R is the scalar uncertainty of the optical flow measurements and I is a 2×2 identity matrix.

Simplified Formulation

The motion of a typical airplane is mostly in the forward direction, that is, the speed of the aircraft is primarily contained in the component, u, while V and w are small. With this idea, assuming that V and w are zero, the formulation is simplified to the following state vector, x, bias state vector, b, input vector, u, optical flow input vectors, d_i, and output vectors, z_i:

$$\mathbf{x} = \begin{bmatrix} u & \phi & \theta & \mathbf{b}^T \end{bmatrix}^T,$$

$$\mathbf{b} = \begin{bmatrix} b_{a_x} & b_p & b_q & b_r & b_L \end{bmatrix}^T,$$

$$\mathbf{u} = \begin{bmatrix} a_x & p & q & r & L \end{bmatrix}^T,$$

$$\mathbf{d}_i = \begin{bmatrix} \mu & v \end{bmatrix}_i^T,$$

$$\mathbf{z}_i = \dot{\mu}_i.$$

$$(14)$$

Note that this simplified formulation removes the V and w states which removes the need for y-axis and z-axis acceleration measurements. Since the yaw state is not contained in any of the state or observation equations it has also been removed. Due to the assumption that V and w are zero, only the x-direction of optical flow is relevant. With these simplifications, the state dynamics become

$$\dot{u} = a_x - g \sin \theta,$$

$$\dot{\phi} = p + q \sin \phi \tan \theta + r \cos \phi \tan \theta,$$

$$\dot{\theta} = q \cos \phi - r \sin \phi.$$

$$(15)$$

The dynamics of the bias states remain the same as in the full formulation except the corresponding bias states for a_y and a_z have been removed. The observation equations from (2) are simplified to be

$$\dot{\mu} = \frac{-u \left(f + v \tan \phi - \mu \tan \theta / \cos \phi \right)}{\eta_z} - fq + rv$$

$$+ \frac{p\mu v}{f} - \frac{q\mu^2}{f}.$$

$$(16)$$

The advantage of considering this simplified formulation is primarily to reduce the computational complexity of the system. The processing of vision data leading to a relatively large number of measurement updates can significantly drive up the computation time of the system, particularly for higher sampling rates. This simplified formulation not only reduces the computation time through a reduction of states, but also significantly reduces the processing and update time for optical flow measurements since only the forward component of flow is used. This formulation could be more practical than the full state formulation for real-time implementation, especially on systems which are limited in onboard computational power due, for example, to cost or size constraints.

Inclusion of a Range State

It is possible to include a state to estimate the range in order to recover the scale of the optical flow images. To determine the dynamics of the range state, the flat ground assumption is used. With this assumption, consider the projection of the range vector onto the Earth-fixed -axis, that is, "down," as shown in Figure 1, by taking the projection through both the roll and pitch angles of the aircraft:

$$z = -L \cos \phi \cos \theta. \tag{17}$$

Here, the negative sign is used because the L coordinate is always positive, while the z coordinate will be negative when the aircraft is above the ground (due to the "down" convention). Taking the derivative with respect to time yields

$$\dot{z} = -\dot{L} \cos \phi \cos \theta + \dot{\phi}L \sin \phi \cos \theta + \dot{\theta}L \cos \phi \sin \theta. \tag{18}$$

Compare this z-velocity equation with that obtained from rotating aircraft body velocity components into the Earthfixed frame:

$$\dot{z} = -u \sin \theta + v \sin \phi \cos \theta + w \cos \phi \cos \theta. \tag{19}$$

Equating these two expressions for z-velocity gives

$$- \dot{L} \cos \phi \cos \theta + \dot{\phi}L \sin \phi \cos \theta + \dot{\theta}L \cos \phi \sin \theta$$
$$= -u \sin \theta + v \sin \phi \cos \theta + w \cos \phi \cos \theta. \tag{20}$$

Simplifying this relationship leads to

$$\dot{L} = \dot{\phi}L \tan \phi + \dot{\theta}L \tan \theta + u \sec \phi \tan \theta - v \tan \phi - w. \tag{21}$$

Substituting in the dynamics for the roll and pitch angles and simplifying leads to the following expression for the range state dynamics:

$$\dot{L} = u \sec \phi \tan \theta - v \tan \phi - w$$
$$+ L \left[p \tan \phi + q \sec \phi \tan \theta \right]. \tag{22}$$

Note that, for level conditions, that is, roll and pitch angles are zero, the equation reduces to

$$\dot{L} = -w \tag{23}$$

which agrees with physical intuition. In order to implement the range state in the simplified formulation, the following expression can be used:

$$\dot{L} = u \sec \phi \tan \theta + L \left[p \tan \phi + q \sec \phi \tan \theta \right]. \tag{24}$$

Information Fusion Algorithm

Due to the nonlinearity, nonadditive noise and numbers of multiple optical flow measurements ranging from 0 to 300 per frame with a mean of 250, the Unscented Information Filter (UIF) [39–41] was selected for the implementation of this algorithm [42]. The advantage of the information filtering framework over Kalman filtering is that redundant information vectors are additive [39–41]; therefore, the time-varying number of outputs obtained from optical flow can easily be handled with relatively low computation, since the coupling between the errors in different optical flow measurements is neglected. The UIF algorithm is summarized as follows [41].

Consider a discrete time nonlinear dynamic system of the form

$$\mathbf{x}_k = \mathbf{f} \left(\mathbf{x}_{k-1}, \mathbf{u}_{k-1} \right) + \mathbf{w}_{k-1}, \tag{25}$$

with measurement equations of the form

$$\mathbf{z}_k = \mathbf{h} \left(\mathbf{x}_k, \mathbf{d}_k \right) + \mathbf{v}_k, \tag{26}$$

where h is the observation function and w and v are the zeromean Gaussian process and measurement noise vectors. At each time step, sigma-points are generated from the prior distribution using

$$\boldsymbol{\chi}_{k-1}$$

$$= \left[\hat{\mathbf{x}}_{k-1} \quad \hat{\mathbf{x}}_{k-1} + \sqrt{N + \lambda} \sqrt{\mathbf{P}_{k-1}} \quad \hat{\mathbf{x}}_{k-1} - \sqrt{N + \lambda} \sqrt{\mathbf{P}_{k-1}} \right], \tag{27}$$

where N is the total number of states and λ is a scaling parameter [42]. Now, the sigma-points are predicted using

$$\boldsymbol{\chi}_{k|k-1}^{(i)} = \mathbf{f} \left(\boldsymbol{\chi}_{k-1}^{(i)}, \mathbf{u}_{k-1} \right), \quad i = 0, 1, \ldots, 2N, \tag{28}$$

where (i) denotes the ith column of a matrix. The a priori statistics are then recovered:

$$\hat{\mathbf{x}}_{k|k-1} = \sum_{i=0}^{2N} \eta_i^m \chi_{k|k-1}^{(i)},$$

$$\mathbf{P}_{k|k-1} = \mathbf{Q}_{k-1}$$

$$+ \sum_{i=0}^{2N} \eta_i^c \left(\chi_{k|k-1}^{(i)} - \hat{\mathbf{x}}_{k|k-1} \right) \left(\chi_{k|k-1}^{(i)} - \hat{\mathbf{x}}_{k|k-1} \right)^T,$$

(23)

where Q is the process noise covariance matrix, and η_i^m and η_i^c are weight vectors [42]. Using these predicted values, the information vector, y, and matrix, Y, are determined:

$$\hat{\mathbf{y}}_{k|k-1} = \mathbf{P}_{k|k-1}^{-1} \hat{\mathbf{x}}_{k|k-1},$$

$$\mathbf{Y}_{k|k-1} = \mathbf{P}_{k|k-1}^{-1}.$$

(30)

For each measurement, that is, each optical flow pair, the output equations are evaluated for each sigma-point, as in

$$\psi_{k|k-1}^{(i,j)} = \mathbf{h} \left(\chi_{k|k-1}^{(i,j)}, \mathbf{d}_k \right),$$

$$i = 0, 1, \ldots, 2N, \quad j = 1, \ldots, n_k,$$

(31)

where ψ denotes an output sigma-point and the superscript (i, j) denotes the ith sigma-point and the jth measurement. The computed observation is then recovered using

$$\hat{\mathbf{z}}_{k|k-1}^{(j)} = \sum_{i=0}^{2N} \eta_i^m \psi_{k|k-1}^{(i)}, \quad j = 1, \ldots, n_k.$$

(32)

Using the computed observation, the cross-covariance is calculated:

$$\mathbf{P}_{k|k-1}^{xy(j)} = \sum_{i=0}^{2N} \eta_i^c \left(\chi_{k|k-1}^{(i)} - \hat{\mathbf{x}}_{k|k-1} \right) \left(\psi_{k|k-1}^{(i)} - \hat{\mathbf{z}}_{k|k-1} \right)^T.$$

(33)

Then the observation sensitivity matrix, , is determined:

$$\mathbf{H}_k^{(j)} = \left[\mathbf{P}_{k|k-1}^{-1} \mathbf{P}_{k|k-1}^{xy(j)} \right]^T, \quad j = 1, \ldots, n_k.$$

(34)

The information contributions can then be calculated:

$$\hat{\mathbf{y}}_k = \hat{\mathbf{y}}_{k|k-1}$$

$$+ \sum_{j=1}^{n_k} \left(\mathbf{H}_k^{(j)}\right)^T \mathbf{R}_k^{-1} \left[\mathbf{z}_k^{(j)} - \mathbf{z}_{k|k-1}^{(j)} + \mathbf{H}_k^{(j)}\hat{\mathbf{x}}_{k|k-1}\right],$$

$$\mathbf{Y}_k = \mathbf{Y}_{k|k-1} + \sum_{j=1}^{n_k} \left(\mathbf{H}_k^{(j)}\right)^T \mathbf{R}_k^{-1}\mathbf{H}_k^{(j)}.$$

$$(35)$$

EXPERIMENTAL SETUP

The research platform used for this study is the West Virginia University (WVU) "Red Phastball" UAV, shown in Figure 2, with a custom GPS/INS data logger mounted inside the aircraft [28, 43]. Some details for this aircraft are provided in Table 1.

Table 1: Details for WVU Phastball aircraft

Property	Value
Length	2.2 m
Wingspan	2.4 m
Takeoff mass	11 kg
Payload mass	3 kg
Propulsion system	Dual 90 mm Ducted Fan Motor
Static thrust	60 N
Fuel capacity	Two 5 Ah LiPo batteries
Cruise speed	30 m/s
Mission duration	6 min

Figure 2: Picture of WVU "Red Phastball" aircraft.

The IMU used in this study is an Analog Devices ADIS-16405 MEMS-based IMU, which includes triaxial accelerometers and rate gyroscopes. Each suite of sensors on the IMU is acquired at 18-bit resolution at 50 Hz over ranges of ±18 g's and ±150 deg/s, respectively. The GPS receiver used in the data logger is a Novatel OEM-V1, which was configured to provide Cartesian position and velocity measurements and solution standard deviations at a rate of 50 Hz, with 1.5 m RMS horizontal position accuracy and 0.03 m/s RMS velocity accuracy. An Optic-Logic RS400 laser range finder was used for range measurement with an approximate accuracy of 1 m and range of 366 m, pointing downward. In addition, a high-quality Goodrich mechanical vertical gyroscope is mounted onboard the UAV to provide pitch and roll measurements to be used as sensor fusion "truth" data, with reported accuracy of within 0.25° of true vertical. The vertical gyroscope measurements were acquired at 16-bit resolution with measurement ranges of ±80 deg for roll and ±60 deg for pitch.

A GoPro Hero video camera is mounted at the center of gravity of the UAV for flight video collection, pointing downwards. The camera was previously calibrated to a focal length of 1141 pixels [29]. Two different sets of flight data were used for this study, each using different camera settings. The first flight used a pixel size of 1920 × 1080 and a sampling rate of 29.97 Hz. The second flight used a pixel size of 1280 × 720 and a sampling rate of 59.94 Hz. All the other sensor data were collected at 50 Hz and resampled to the camera time for postflight validation after manual synchronization.

EXPERIMENTAL RESULTS

Flight Data

Two sets of flight data from the WVU "Red Phastball" aircraft were used in this study. Each flight consists of approximately 5 minutes of flight. The top-down flight trajectories from these two data sets are overlaid on a Google Earth image of the flight test location in Figure 3. Six different unique markers have been placed in Figure 3 in order to identify specific points along the trajectory. These markers will be used in future figures in order to synchronize the presentation of data.

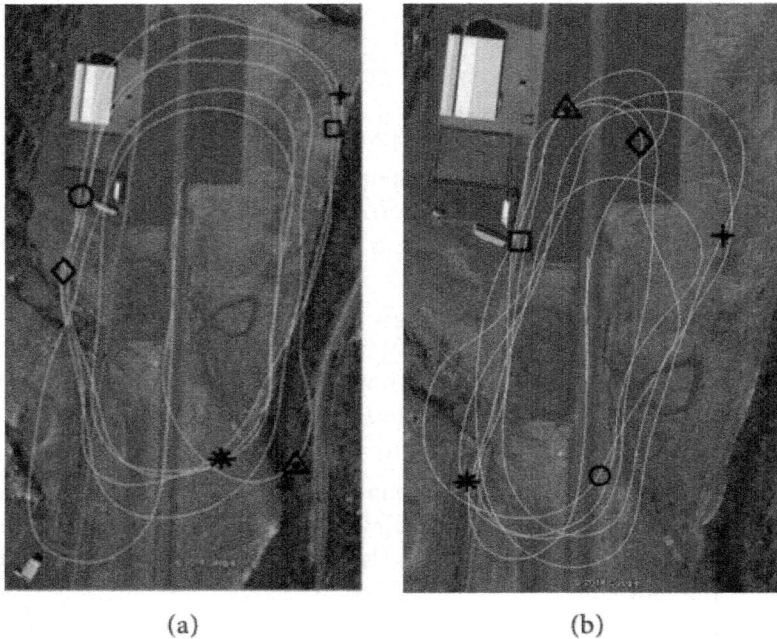

(a)   (b)

Figure 3: Flight trajectories for Flight #1 (a) and Flight #2 (b), © 2014 Google.

Selection of Noise Assumptions for Optical Flow Measurements

Since the noise properties of the IMU have been established from previous work [33], only the characteristics of the uncertainty in the laser range and optical flow measurements need to be determined. The uncertainty in the laser range finder measurement is modeled as 1 m zero-mean Gaussian noise, based on the manufacturer's reported accuracy of the sensor. The optical flow errors are a bit more difficult to model. Due to this difficulty, different assumptions of the optical flow uncertainty were considered. Using both sets of the flight data, the full state UIF was executed for each assumption of optical flow uncertainty. To evaluate the performance of the filter, the speed measurements were compared with reference measurements from GPS which have been mapped into the aircraft frame using roll and pitch measurements from the vertical gyroscope and approximating the yaw from the heading as determined by GPS. The roll and pitch estimates were compared with the measurements from the vertical gyroscope. Due to the possibility of alignment errors, only standard deviation of error was considered. Each of these errors was calculated for each set of flight data, and the results are offered in Figure 4.

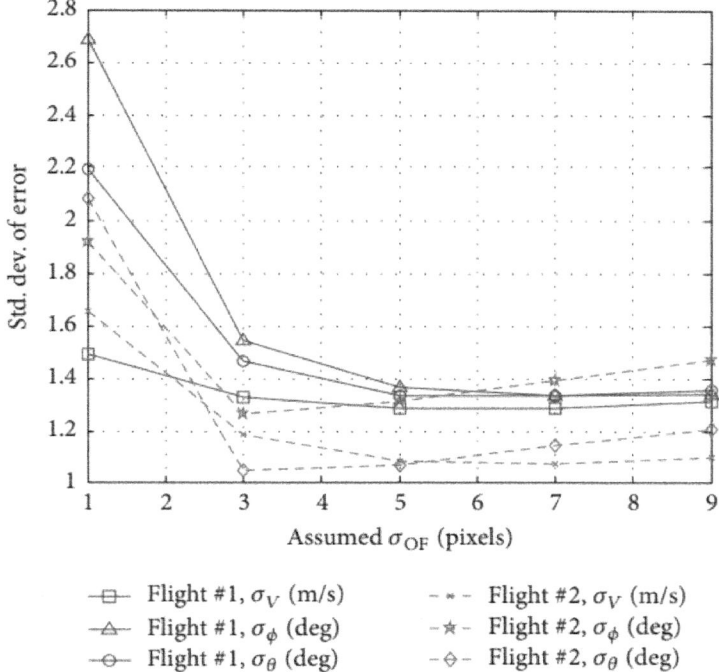

Figure 4: Comparison of errors for different assumed optical flow uncertainties.

Figure 4 shows how changing the assumption on the optical flow uncertainty affects the estimation performance of the total ground speed, roll angle, and pitch angle. The relatively flat region in Figure 4 for assumed optical flow standard deviations from approximately 3 to 9 pixels indicates that this formulation is relatively insensitive to tuning of these optical flow errors. It is also interesting to note in Figure 4 that Flight #1 and Flight #2 have optimum performance at different values of R. This however makes sense, as Flight #2 has twice the frame rate as Flight #1; therefore, the assumed noise characteristics should be one half that of Flight #1. From Figure 4, the optical flow uncertainties were selected to be R=52 pixels2 for Flight #1 and $R = 2.52$ pixels2 for Flight #2.

Full State Formulation Estimation Results

Using each set of flight data, the full state formulation using UIF was executed. The estimated components of velocity are shown for Flight #1 in Figure 5 and for Flight #2 in Figure 6. These estimates from the UIF are offered with respect to comparable reference values from GPS, which were mapped into the aircraft frame using roll and pitch measurements from the vertical gyroscope and approximating the yaw angle with the heading angle obtained from GPS. From each of these figures, the following observations can be made.The forward velocity, u, is reasonably captured by the estimation. The lateral velocity, V, and vertical velocity, w, however, demonstrate somewhat poor results. This does however make sense, as the primary direction of flight is forward, thus resulting in good observability characteristics in the optical flow in the forward direction, while the signal-to-noise ratio (SNR) for the lateral and vertical directions remains small for most typical flight conditions. However, since these lateral and vertical components are only a small portion of the total velocity, the total speed can be reasonably approximated by this technique. The total speed estimates are shown in Figure 7 for Flight #1 with GPS reference.

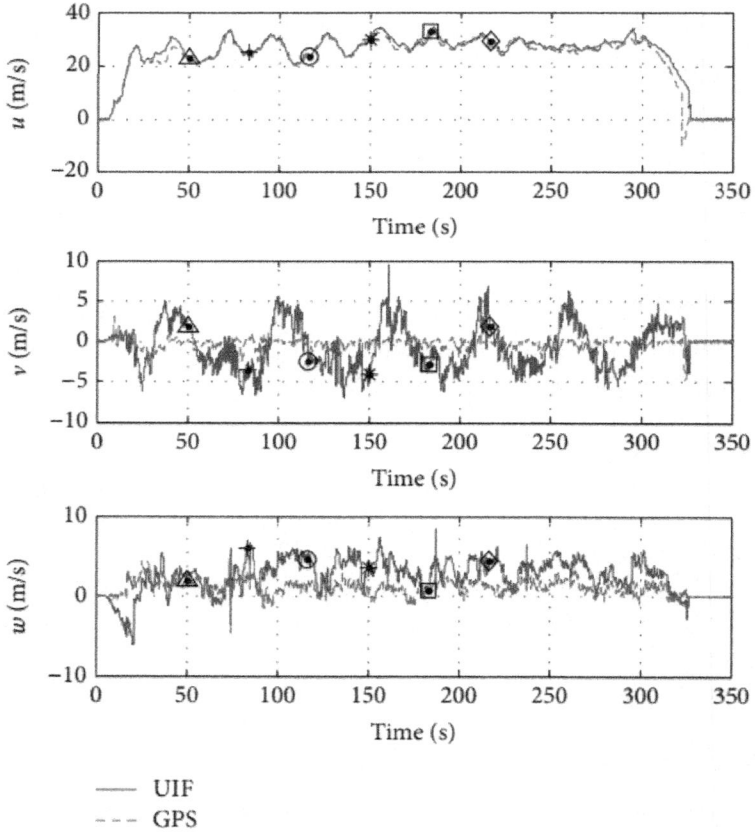

Figure 5: Estimated velocity components for Flight #1.

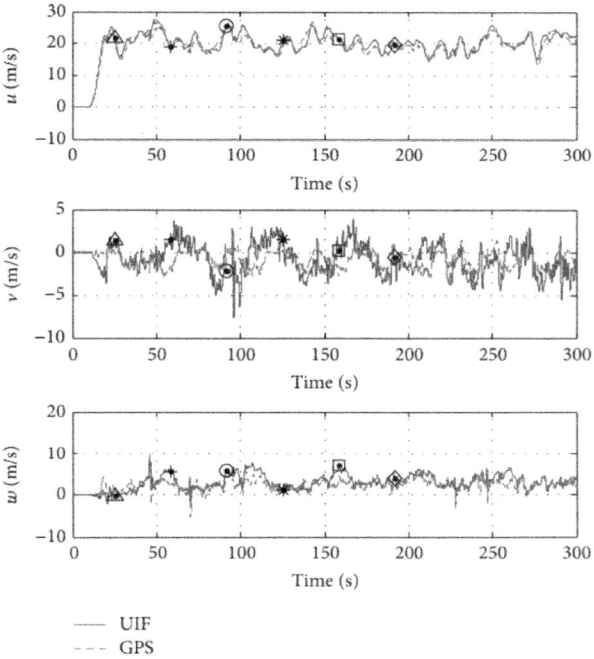

Figure 6: Estimated velocity components for Flight #2.

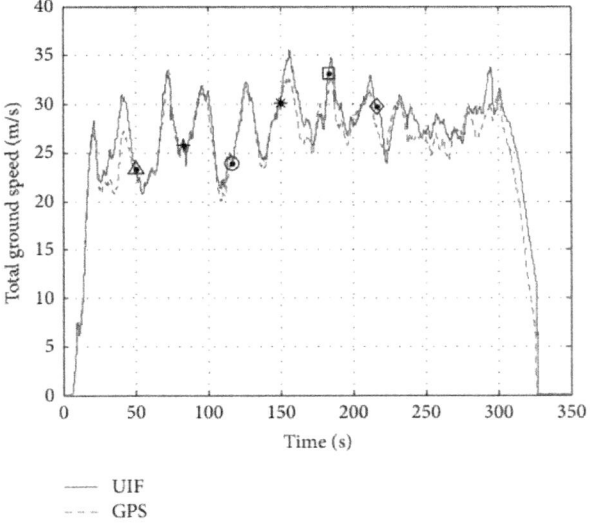

Figure 7: Estimated total speed for Flight #1.

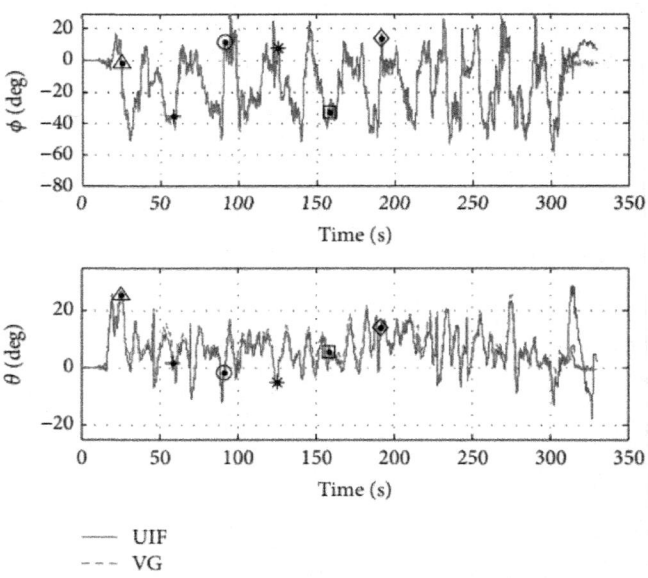

Figure 8: Roll and pitch estimation results for Flight #2.

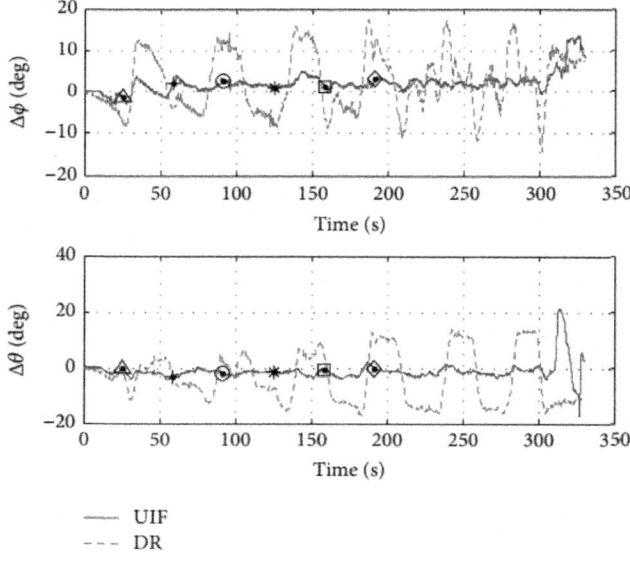

Figure 9: Roll and pitch estimation errors as compared to dead reckoning (DR) for Flight #2.

The attitude estimates for the roll and pitch angles are compared with the vertical gyroscope measurements as a reference, as shown in Figure 8. In order to demonstrate the effectiveness of this method in regulating the drift in attitude estimates that occurs with dead reckoning, the estimation errors from the UIF are compared with the errors obtained from dead reckoning attitude estimation. These roll and pitch errors are offered in Figure 9for Flight #2. Figure 9 demonstrates the effectiveness of the UIF in regulating the attitude errors from dead reckoning.

In order to quantify the estimation results, the mean absolute error and standard deviation of error of the estimates are calculated for the velocity components with respect to the GPS reference and also for the roll and pitch angles with respect to the vertical gyroscope reference. These statistical results are provided in Table 2 for Flight #1 and Table 3 for Flight #2, where v is the total airspeed as determined by

$$V = \sqrt{u^2 + v^2 + w^2}. \tag{36}$$

It is shown in Tables 2 and 3 that reasonable errors are obtained in both sets of flight data for the velocity and attitude of the aircraft. Larger errors are noted in particular for the lateral velocity state, V, which is due to observability issues in the optical flow. Note that mean errors in the roll and pitch estimation could be due to misalignment between the vertical gyroscope, IMU, and video camera. The attitude estimation accuracy is reported in Tables 2 and 3 similar to the reported accuracy of loosely coupled GPS/INS attitude estimation using similar flight data [43].

Table 2: Flight #1 error statistics for estimated states

Estimated state	Mean abs.	Standard deviation	Units
u	1.0644	1.2667	m/s
v	2.4699	2.7719	m/s
w	2.1554	1.6554	m/s
V	1.2466	1.2858	m/s
ϕ	1.1553	1.3668	deg
θ	2.1752	1.3339	deg

Table 3: Flight #2 error statistics for estimated states

Estimated state	Mean abs.	Standard deviation	Units
u	1.1299	1.2663	m/s
v	1.5184	1.8649	m/s
w	1.1324	1.3453	m/s
V	1.2087	1.2535	m/s
ϕ	1.8818	1.2782	deg
θ	1.6646	1.1049	deg

Simplified Formulation Estimation Results

Since it was observed in the full state formulation results that the lateral and vertical estimates were small, the simplified formulation was implemented in order to investigate the feasibility of a simplified version of the filter that estimates only the forward velocity component and assumes the lateral and vertical components are zero. The forward velocity, u, for Flight #1 is offered in Figure 10, while the roll and pitch errors with respect to the vertical gyroscope measurement are offered in Figure 11 for the UIF and dead reckoning (DR). Additionally, the mean absolute error and standard deviation of error for these terms are provided in Table 4 for Flight #1 and Table 5 for Flight #2.

Table 4: Simplified formulation error statistics for Flight #1

Estimated state	Mean abs.	Standard deviation	Units
u	1.2283	1.5730	m/s
ϕ	1.9901	2.2922	deg
θ	1.9002	2.1881	deg

Table 5: Simplified formulation error statistics for Flight #2

Estimated state	Mean abs.	Standard deviation	Units
u	1.3073	1.5775	m/s
ϕ	2.8402	3.5426	deg
θ	2.0286	2.4691	deg

It is shown in Tables 4 and 5 that the simplified formulation results in significantly higher attitude estimation errors with respect to the full state formulation. These increased attitude errors are likely due to the assumption that lateral and vertical velocity components are zero. To investigate this possible correlation, the roll and pitch errors are shown in Figure 12 with the magnitude of the lateral and vertical velocity as determined from GPS for a 50-second segment of flight data which includes takeoff. Figure 12 shows that there is some correlation between the attitude estimation errors and the lateral and vertical velocity, though it is not the only source of error for these estimates.

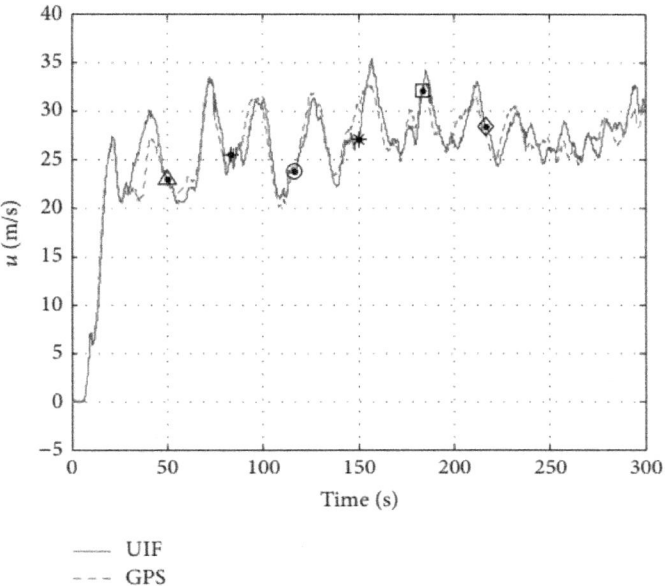

Figure 10: Simplified formulation speed estimation results for Flight #1.

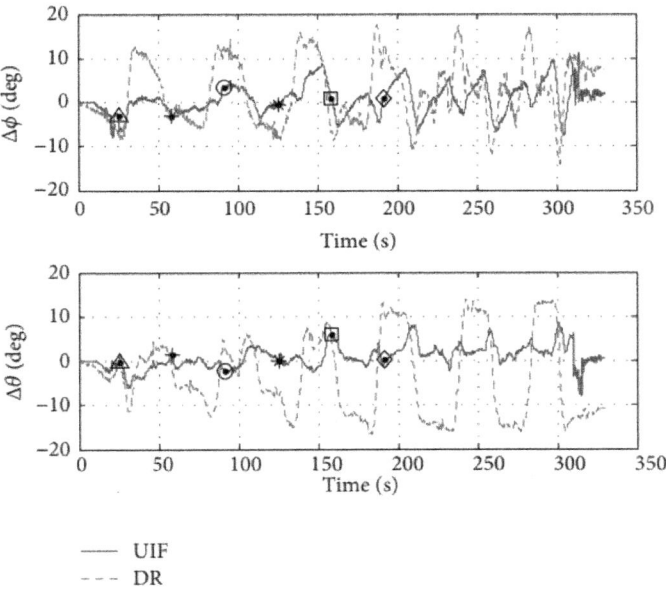

Figure 11: Simplified formulation attitude estimates with dead reckoning (DR) for Flight #2.

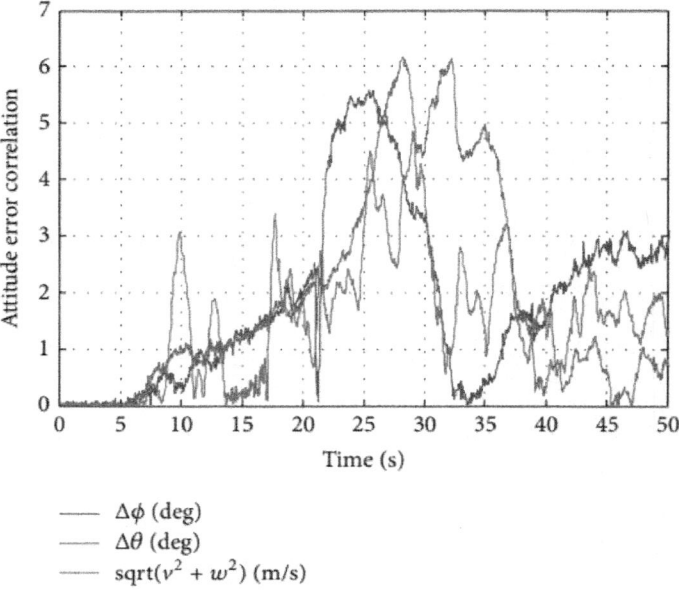

Figure 12: Comparison of attitude estimation errors with respect to lateral and vertical velocity.

Results Using Range State

The results for each flight for both the full state formulation and simplified formulation were recalculated with the addition of the range state. The statistical results for these tests are offered in Tables 6–9.

Table 6: Flight #1 error statistics for estimated states with range state

Estimated state	Mean abs.	Standard deviation	Units
u	1.1330	1.3699	m/s
v	2.4387	2.7369	m/s
w	2.1210	1.5998	m/s
V	1.3041	1.3852	m/s
ϕ	1.1599	1.4084	deg
θ	2.1767	1.4064	deg

Table 7: Flight #2 error statistics for estimated states with range state

Estimated state	Mean abs.	Standard deviation	Units
u	1.1444	1.2937	m/s
v	1.5122	1.8529	m/s
w	1.1154	1.3268	m/s
V	1.2112	1.2761	m/s
ϕ	1.8897	1.2845	deg
θ	1.6609	1.1152	deg

Table 8: Simplified formulation error statistics for Flight #1 with range state

Estimated state	Mean abs.	Standard deviation	Units
u	1.2919	1.6256	m/s
ϕ	2.0621	2.3746	deg
θ	1.9277	2.2713	deg

Table 9: Simplified formulation error statistics for Flight #2 with range state.

Estimated state	Mean abs.	Standard deviation	Units
u	1.3356	1.6016	m/s
ϕ	2.8748	3.5462	deg
θ	2.0303	2.4876	deg

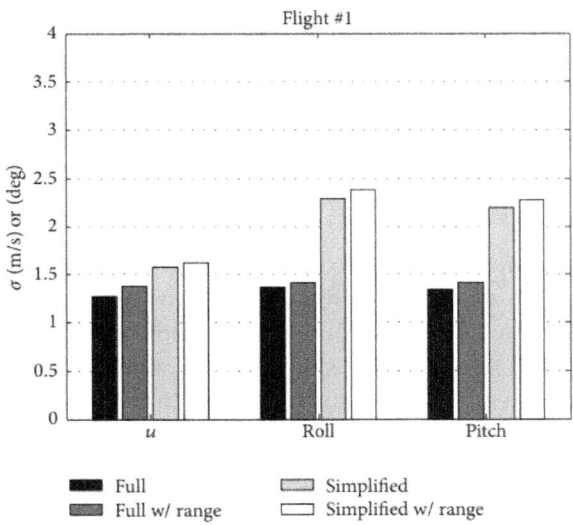

Figure 13: Graphical comparison of standard deviation of error for Flight #1.

In order to compare the results from the different cases, the standard deviation of error is shown graphically for Flight #1 in Figure 13 and Flight #2 in Figure 14. It is shown in Figures 13 and 14 that the simplified formulation offers poorer estimation performance as expected, particularly for the attitude estimates. The addition of the range state does not affect the performance significantly.

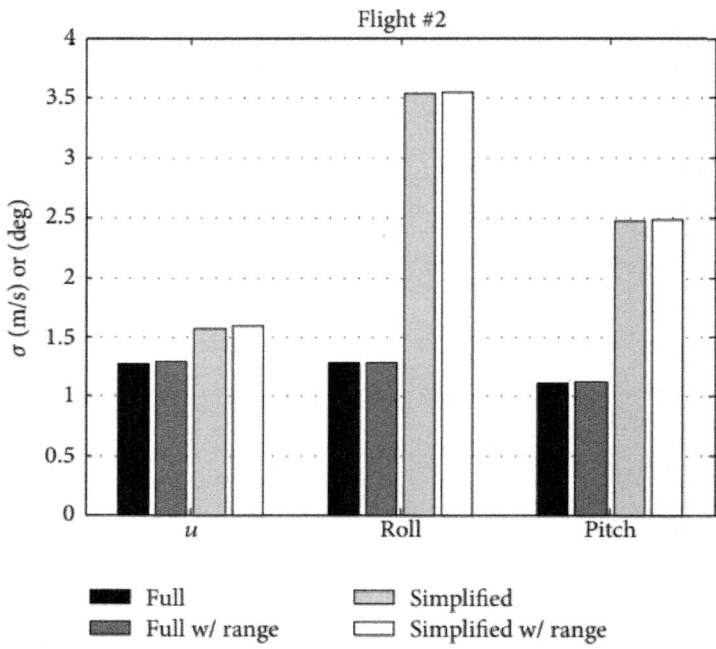

Figure 14: Graphical comparison of standard deviation of error for Flight #2.

CONCLUSIONS

This paper presented vision-aided inertial navigation techniques which do not rely upon GPS using UAV flight data. Two different formulations were presented, a full state estimation formulation which captures the aircraft ground velocity vector and attitude and a simplified formulation which assumes all of the aircraft velocity is in the forward direction. Both formulations were shown to be effective in regulating the INS drift. Additionally, a state was included in each formulation in order to estimate the distance between the image center and the aircraft. The full state formulation was shown to be effective in estimating aircraft ground velocity to within 1.3 m/s and regulating attitude angles within 1.4 degrees standard deviation of error for both sets of flight data.

ACKNOWLEDGMENTS

This work was supported in part by NASA Grant no. NNX12AM56A, Kansas NASA EPSCoR PDG grant, and SRI grant.

REFERENCES

1.   C. Li, L. Shen, H.-B. Wang, and T. Lei, "The research on unmanned aerial vehicle remote sensing and its applications," in Proceedings of the IEEE International Conference on Advanced Computer Control (ICACC '10), pp. 644–647, Shenyang, China, March 2010.

2.   J. Calise and R. T. Rysdyk, "Nonlinear adaptive flight control using neural networks," IEEE Control Systems Magazine, vol. 18, no. 6, pp. 14–24, 1998.

3.   M. S. Grewal, L. R. Weill, and A. P. Andrew, Global Positioning, Inertial Navigation & Integration, Wiley, New York, NY, USA, 2nd edition, 2007.

4.   J. L. Crassídís, "Sigma-point Kalman filtering for integrated GPS and inertial navigation," in Proceedings of the AIAA Guidance, Navigation, and Control Conference, pp. 1981–2004, San Francisco, Calif, USA, August 2005.

5.   J. N. Gross, Y. Gu, M. B. Rhudy, S. Gururajan, and M. R. Napolitano, "Flight-test evaluation of sensor fusion algorithms for attitude estimation," IEEE Transactions on Aerospace and Electronic Systems, vol. 48, no. 3, pp. 2128–2139, 2012.

6.   M. V. Srinivasan, "Honeybees as a model for the study of visually guided flight, navigation, and biologically inspired robotics," Physiological Reviews, vol. 91, no. 2, pp. 413–460, 2011.

7.   P. S. Bhagavatula, C. Claudianos, M. R. Ibbotson, and M. V. Srinivasan, "Optic flow cues guide flight in birds," Current Biology, vol. 21, no. 21, pp. 1794–1799, 2011.

8.   N. Franceschini, "Visual guidance based on optic flow: a biorobotic approach," Journal of Physiology Paris, vol. 98, no. 1-3, pp. 281–292, 2004.

9.   P. Corke, J. Lobo, and J. Dias, "An introduction to inertial and visual sensing," The International Journal of Robotics Research, vol. 26, no. 6, pp. 519–535, 2007.

10.   D. G. Lowe, "Distinctive image features from scale-invariant keypoints," International Journal of Computer Vision, vol. 60, no. 2, pp. 91–110, 2004.

11. M. V. Srinivasan, S. Thurrowgood, and D. Soccol, "Competent vision and navigation systems," IEEE Robotics & Automation Magazine, vol. 16, no. 3, pp. 59–71, 2009.

12. J. Conroy, G. Gremillion, B. Ranganathan, and J. S. Humbert, "Implementation of wide-field integration of optic flow for autonomous quadrotor navigation," Autonomous Robots, vol. 27, no. 3, pp. 189–198, 2009.

13. F. Kendoul, I. Fantoni, and K. Nonami, "Optic flow-based vision system for autonomous 3D localization and control of small aerial vehicles," Robotics and Autonomous Systems, vol. 57, no. 6-7, pp. 591–602, 2009.

14. Beyeler, J.-C. Zufferey, and D. Floreano, "optiPilot: control of take-off and landing using optic flow," inProceedings of the European Micro Air Vehicle Conference and Competition (EMAV ‹09), Delft, The Netherland, September 2009.

15. M. M. Veth and J. Raquet, "Fusion of low-cost imaging and inertial sensors for navigation," inProceedings of the 19th International Technical Meeting of the Satellite Division Institute of Navigation (ION GNSS ‹06), pp. 1093–1103, Fort Worth, Tex, USA, September 2006.

16. D. Dusha, W. Boles, and R. Walker, "Fixed-wing attitude estimation using computer vision based horizon detection," in Proceedings of the International Unmanned Air Vehicle Systems Conference, April 2007.

17. D. Dusha, W. Boles, and R. Walker, "Attitude estimation for a fixed-wing aircraft using horizon detection and optical flow," in Proceedings of the 9th Biennial Conference of the Australian Pattern Recognition Society on Digital Image Computing Techniques and Applications (DICTA ‹07), pp. 485–492, IEEE, Glenelg, Australia, December 2007.

18. S. Weiss, M. W. Achtelik, S. Lynen, M. Chli, and R. Siegwart, "Real-time onboard visual-inertial state estimation and self-calibration of MAVs in unknown environments," in Proceedings of the IEEE International Conference on Robotics and Automation (ICRA ‹12), May 2012.

19. M. Dille, B. Grocholsky, and S. Singh, "Outdoor downward-facing optical flow odometry with commodity sensors," in Field and Service Robotics: Results of the 7th International Conference, vol. 62 ofSpringer Tracts in Advanced Robotics, pp. 183–193, 2010.

20. S. I. Roumeliotis, A. E. Johnson, and J. F. Montgomery, "Augmenting inertial navigation with image-based motion estimation," in Proceedings of the IEEE International Conference on Robotics & Automation, pp. 4326–4333, Washington, DC, USA, May 2002.

21. M. Achtelik, M. Achtelik, S. Weiss, and R. Siegwart, "Onboard IMU and monocular vision based control for MAVs in unknown in- and outdoor environments," in Proceedings of the IEEE International Conference on Robotics and Automation (ICRA ‹11), pp. 3056–3063, Shanghai, China, May 2011.

22. P.-J. Bristeau, F. Callou, D. Vissière, and N. Petit, "The navigation and control technology inside the AR.Drone micro UAV," in Proceedings of the 18th IFAC World Congress, pp. 1477–1484, Milano, Italy, September 2011.

23. H. Chao, Y. Gu, and M. Napolitano, "A survey of optical flow techniques for robotics navigation applications," Journal of Intelligent and Robotic Systems: Theory and Applications, vol. 73, no. 1–4, pp. 361–372, 2014.

24. W. Ding, J. Wang, S. Han et al., "Adding optical flow into the GPS/INS integration for UAV navigation," in Proceedings of the International Global Navigation Satellite Systems Society IGNSS Symposium, Surfers Paradise, Australia, 2009.

25. H. Romero, S. Salazar, and R. Lozano, "Real-time stabilization of an eight-rotor UAV using optical flow,"IEEE Transactions on Systems, Man, and Cybernetics—Part C: Applications and Reviews, vol. 42, no. 6, pp. 1752–1762, 2012.

26. V. Grabe, H. H. Bulthoff, and P. R. Giordano, "A comparison of scale estimation schemes for a quadrotor UAV based on optical flow and IMU measurements," in Proceedings of the 26th IEEE/RSJ International Conference on Intelligent Robots and Systems (IROS ‹13), pp. 5193–5200, IEEE, Tokyo, Japan, November 2013.

27. Martinelli, "Vision and IMU data fusion: closed-form solutions for attitude, speed, absolute scale, and bias determination,"IEEE Transactions on Robotics, vol. 28, no. 1, pp. 44–60, 2012.

28. M. B. Rhudy, H. Chao, and Y. Gu, "Wide-field optical flow aided inertial navigation for unmanned aerial vehicles," in Proceedings of the IEEE/RSJ International Conference on Intelligent Robots and Systems (IROS ‹14), pp. 674–679, Chicago, Ill, USA, September 2014.

29. H. Chao, Y. Gu, J. Gross, G. Guo, M. L. Fravolini, and M. R. Napolitano, "A comparative study of optical flow and traditional sensors in UAV navigation," in Proceedings of the 1st American Control Conference (ACC ‹13), pp. 3858–3863, Washington, DC, USA, June 2013.

30. R. C. Hibbeler, Engineering Mechanics: Dynamics, Prentice Hall, 10th edition, 2004.

31. V. Klein and E. A. Morelli, Aircraft System Identification: Theory and Practice, American Institute of Aeronautics and Astronautics, Reston, Va, USA, 2006.

32. J. Roskam, Airplane Flight Dynamics and Automatic Flight Controls, DARcorporation, Lawrence, Kan, USA, 2003.

33. J. N. Gross, Y. Gu, M. Rhudy, F. J. Barchesky, and M. R. Napolitano, "On-line modeling and calibration of low-cost navigation sensors," in Proceedings of the AIAA Modeling and Simulation Technologies Conference, pp. 298–311, Portland, Ore, USA, August 2011.

34. D. W. Allan, "Statistics of atomic frequency standards," Proceedings of the IEEE, vol. 54, no. 2, pp. 221–230, 1966.

35. X. Zhiqiang and D. Gebre-Egziabher, "Modeling and bounding low cost inertial sensor errors," inProceedings of the IEEE/ION Position, Location and Navigation Symposium (PLANS ‹08), pp. 1122–1132, IEEE, Monterey, Calif, USA, May 2008.

36. Z. Xing, Over-bounding integrated INS/GNSS output errors [Ph.D. dissertation], The University of Minnesota, Minneapolis, Minn, USA, 2010.

37. F. L. Lewis and V. L. Syrmos, Optimal Control, Wiley, New York, NY, USA, 2nd edition, 1995.

38. M. Mammarella, G. Campa, M. L. Fravolini, Y. Gu, B. Seanor, and M. R. Napolitano, "A comparison of optical flow algorithms for real time aircraft guidance and navigation," in Proceedings of the AIAA Guidance, Navigation and Control Conference and Exhibit, Honolulu, Hawaii, USA, August 2008.

39. G. O. Mutambara, Decentralized Estimation and Control for Multisensor Systems, CRC Press, Washington, DC, USA, 1998.

40. T. Vercauteren and X. Wang, "Decentralized sigma-point information filters for target tracking in collaborative sensor networks," IEEE Transactions on Signal Processing, vol. 53, no. 8, part 2, pp. 2997–3009, 2005.

41. D.-J. Lee, "Unscented information filtering for distributed estimation and multiple sensor fusion," inProceedings of the AIAA Guidance, Navigation and Control Conference and Exhibit, Honolulu, Hawaii, USA, 2008.

42. M. Rhudy and Y. Gu, "Understanding nonlinear Kalman filters," in Interactive Robotics Letters, West Virginia University, 2013, http://www2.statler.wvu.edu/~irl/page13.html.

43.   M. Rhudy, J. Gross, Y. Gu, and M. R. Napolitano, "Fusion of GPS and redundant IMU data for attitude estimation," in Proceedings of the AIAA Guidance, Navigation, and Control Conference, Minneapolis, Minn, USA, August 2012.

CITATION

CHAPTER 1

G. Barceló, "Analysis of Dynamics Fields in Noninertial Systems," *World Journal of Mechanics*, Vol. 2 No. 3, 2012, pp. 175-180. doi: 10.4236/wjm.2012.23021.

CHAPTER 2

Y. Kubota and O. Mochizuki, "Role on Moment of Inertia and Vortex Dynamics for a Thin Rotating Plate," *World Journal of Mechanics*, Vol. 3 No. 6, 2013, pp. 270-276. doi: 10.4236/wjm.2013.36028.

CHAPTER 3

Liu, C. , Tanaka, Y. and Fujimoto, Y. (2015) Viscosity Transient Phenomenon during Drop Impact Testing and Its Simple Dynamics Model. *World Journal of Mechanics*, 5, 33-41. doi: 10.4236/wjm.2015.53004.

CHAPTER 4

Siddique, J. , Landis, F. and Mohyuddin, M. (2014) Dynamics of Drainage of Power-Law Liquid into a Deformable Porous Material. *Open Journal of Fluid Dynamics*, 4, 403-414. doi: 10.4236/ojfd.2014.44030.

CHAPTER 5

Bacchieri, A. (2014) The Light as Composed of Longitudinal-Extended Elastic Particles Obeying to the Laws of Newtonian Mechanics. *Journal of Modern Physics*, 5, 884-899. doi: 10.4236/jmp.2014.59092.

CHAPTER 6

Carranza, D. and Mendoza, S. (2015) Modified Newtonian Dynamics as an Entropic Force. *Journal of Modern Physics*, 6, 786-793. doi: 10.4236/jmp.2015.66084.

CHAPTER 7

Arbab, A. (2012) The Generalized Newton's Law of Gravitation Versus the General Theory of Relativity. *Journal of Modern Physics*, **3**, 1231-1235. doi: 10.4236/jmp.2012.329159.

CHAPTER 8

E. Goulart, "Mimicking General Relativity with Newtonian Dynamics," ISRN Mathematical Physics, vol. 2012, Article ID 260951, 15 pages, 2012. doi:10.5402/2012/260951

CHAPTER 9

Michele Betti, Paolo Biagini, and Luca Facchini, "A Hybrid Approach for the Random Dynamics of Uncertain Systems under Stochastic Loading," Mathematical Problems in Engineering, vol. 2011, Article ID 213094, 30 pages, 2011. doi:10.1155/2011/213094

CHAPTER 10

T. G. Ritto;C. Soize;and Rubens Sampaio, Stochastic dynamics of a drill-string with uncertain weight-on-hook, http://dx.doi.org/10.1590/S1678-58782010000300008

CHAPTER 11

Aly Mousaad Aly, Alberto Zasso, and Ferruccio Resta, "Dynamics and Control of High-Rise Buildings under Multidirectional Wind Loads," Smart Materials Research, vol. 2011, Article ID 549621, 15 pages, 2011. doi:10.1155/2011/549621

CHAPTER 12

Z. Q. Zhang, D. K. Ward, Y. Xue, H. W. Zhang, and M. F. Horstemeyer, "Interfacial Characteristics of Carbon Nanotube-Polyethylene Composites Using Molecular Dynamics Simulations," ISRN Materials Science, vol. 2011, Article ID 145042, 10 pages, 2011. doi:10.5402/2011/145042

CHAPTER 13

Matthew B. Rhudy, Yu Gu, Haiyang Chao, and Jason N. Gross, "Unmanned Aerial Vehicle Navigation Using Wide-Field Optical Flow and Inertial Sensors," Journal of Robotics, vol. 2015, Article ID 251379, 12 pages, 2015. doi:10.1155/2015/251379.

INDEX